SpringerBriefs in Applied Sciences and Technology

Manufacturing and Surface Engineering

T0185158

Series Editor
J. Paulo Davim

For further volumes:
http://www.springer.com/series/10623

Angelos P. Markopoulos

Finite Element Method
in Machining Processes

 Springer

Angelos P. Markopoulos
Laboratory of Manufacturing Technology
National Technical University of Athens
Athens
Greece

ISBN 978-1-4471-4329-1   ISBN 978-1-4471-4330-7   (eBook)
DOI 10.1007/978-1-4471-4330-7
Springer London Heidelberg New York Dordrecht

Library of Congress Control Number: 2012941280

Printed on acid-free paper

Springer is part of Springer Science+Business Media (www.springer.com)

Preface

The finite element method is a powerful tool with applications in many industrial sectors. Manufacturing and especially machining is not an exception. Since the early 1970s it has provided valuable information on fundamental understanding of the material removal process but more importantly predictive models that can provide reliable results on many machining parameters. As a modeling method it has proven superior and by far more versatile than any other. The vast number of publications, pertaining to machining, and finite element analysis, proves it.

This Springer Brief aims to provide information on the modeling of machining processes by the finite element method. In Chap. 1 an introduction to machining and especially metal cutting is provided. Some basic features of turning, milling, and drilling are discussed and the terminology that is used throughout the book is introduced. Chapter 2 gives a description of orthogonal and oblique cutting, two schemes very popular among the machining researchers. A discussion on analytical modeling of machining is provided in this chapter. Shear plane, Slip-line field, and shear zone models are discussed. This is a helpful introduction to machining modeling. Chapter 3 is dedicated to finite element method. The features to be implemented in a FEM model are described and the most popular approaches are discussed. In this chapter, numerical formulation, mesh, elements, boundary conditions, and contact considerations are analyzed. Furthermore, material and friction modeling are considered and a discussion on chip separation and breakage criteria and adaptive meshing is provided. From all these numerical considerations and the research conducted so far it may be said that there is not a single solution that is acceptable on how the "perfect model" looks like. The search goes on with new ideas and better tools. The researchers' arsenal stores better understanding of machining, more powerful computers, and special software. At the end of this Chapter a bibliographical review is provided along with a brief presentation of commercial FEM software. Chapter 4 is the part of the book where examples of finite element models are given. Three areas, namely high speed machining, 3D modeling, and micromachining, are selected. For each area, a discussion on the work done so far and how the models overcome problems that may arise is provided. Chapter 5 is the last one, where the modeling techniques for other machining operations are described. More specifically, grinding is considered here and a paragraph for non-conventional machining and machining of composite materials is squeezed in. Furthermore, soft computing techniques, molecular dynamics, and meshless methods for machining are presented.

The book is by no means complete, in the sense that for every topic included a lot more can be added. The reader who starts now to get acquainted with FEM models of machining can raise his awareness of what lies ahead of him. The experienced user may review the advancements through all the past years and get new ideas to move on or use it as a reference book.

I would like to thank Professor J. Paulo Davim, Editor of SpringerBriefs Series in Manufacturing and Surface Engineering for his invitation and the trust he put in me to accomplish the task of writing this book, and Professor D. E. Manolakos for his support and valuable advice. I would also like to thank Miss. Quinn from Springer for her assistance and prompt answers. Finally, I thank my family and especially my wife for being so patient with me.

Angelos P. Markopoulos

Contents

Chapter 1
Machining Processes

1.1 Introduction

Surveys indicate that 15 % of all mechanical components value, manufactured in the world, comes from machining operations and that annual expenditure on machine tools and cutting tools are several billion € for industrially developed countries [1, 2]. Although the available data are from some years ago, increasing trends are indicated and are likely to be sustained until today. If labor, machinery, tools and materials costs, social impact from employment in machining related jobs and technological developments becoming available from machining advances are considered, then the importance of machining and its impact on today's industry and society is obvious.

Trends in manufacturing technology are driven by two very important factors, which are closely interconnected, namely better quality and reduced cost. Modern industry strives for products with dimensional and form accuracy and low surface roughness at acceptable cost; an extreme paradigm being micromachining of miniaturized components. From an economic point of view, machining cost reduction achieved through the increase of material removal rate and tool life without compromising surface integrity even for hard-to-machine materials is highly desirable, e.g. turning of hardened steels by CBN tools at increased speeds or as it is usually referred in the literature High Speed Hard Turning. Understanding chip formation mechanisms and predicting cutting forces are of the greatest importance on realizing both the above goals and one way to achieve this, probably the most used one, is modeling.

Machining of metals, although is one of the oldest and very important manufacturing process, has been subjected to systematic study for a little more than a century. Almost for the second half of this time period, studying of metal machining is accompanied by modeling methods. The initial objective of studying and modeling metal machining was to provide a theory which, without any experimental

A. P. Markopoulos, *Finite Element Method in Machining Processes*, SpringerBriefs in Manufacturing and Surface Engineering, DOI: 10.1007/978-1-4471-4330-7_1, © The Author(s) 2013

work, would enable researchers to predict cutting performance and thus solve practical problems confronted in industry. The first analytical models set the basis for more advanced methods developed later in the course of time and when the tools for realistic computational cost and analysis time became available with computer advances. Analytical models supported by metal cutting mechanics and with simplifying assumptions began publishing around 1900s. However, it was not until the 1950s that modeling of machining became a key tool used for understanding the mechanisms of material removing process as well as predicting their performance.

In the early 1970s some pioneering works on machining modeling with the Finite Element Method (FEM) begun to find their way in scientific journals. Over the years and with the increase of computer power as well as the existence of commercial FEM software, this method has proved to be the favorite modeling tool for researchers of the field. This is established by the vast number of publications on this subject as well as the modeling novelties introduced and used, even by the fact that software dedicated solely for the purpose of modeling machining operations exist; more details on the above subjects only marginally discussed here will be presented in Chaps. 3 and 4. Finite element models are used today for gaining knowledge on fundamental aspect of material removing mechanisms but more importantly for their ability to predict important parameters such as cutting forces, temperatures, stresses etc. essential for the prediction of the process outcome, the quality of the final product and in a timely and inexpensive way. The requirements for performing such a task are many; theoretical background, manufacturing experience, accurate data and knowledge on modeling are supplies for building a model and interpreting its results.

In this Book an effort to provide a guide on the modeling of machining is attempted. Firstly, machining processes and some important features such as basic terminology and tool geometry that are important for modeling are discussed in Chap. 1. In Chap. 2 an analysis on some benefits as well as limitations of modeling in general and the requirements of machining modeling in particular is briefly argued. Then, in Chap. 3, some important features of mechanics of machining and analytical modeling attempts are described. The most extended part of this book is the subject of FEM modeling of machining which includes various discussions on FEM formulation, material modeling, friction and FEM software among others. Chapter 4 includes case studies to exemplify the use FEM in machining and in various processes in 2D and 3D problems as well as literature review and further reading suggestion, wherever this is needed. The results of the analyses are discussed and useful conclusions are drawn. Some special cases of state-of-the-art machining processes and their peculiarities are also provided. Finally, in Chap. 5, some other modeling methods, besides FEM, that are currently used for modeling and simulating machining are discussed.

1.2 Machining Processes: Metal Cutting

By the term *machining*, processes that shape parts by removing unwanted material, are described. The part being machined, usually called workpiece, can be metallic or non-metallic, i.e. polymer, wood, ceramic or even composite, however, machining of metals will be discussed hereafter. Unwanted material is carried away from the workpiece usually in the form of a chip; evaporation or ablation may take place in some machining operations.

The more narrow term *cutting* is used to describe the formation of a chip via the interaction of a tool in the form of a wedge with the surface of the workpiece, given that there is a relative movement between them. These machining operations include turning, milling, drilling and boring among others and are usually referred as traditional machining processes. Abrasive processes such as grinding are also part of cutting processes of great importance in contemporary industry. Other non-traditional machining operations that may or may not include physical contact between cutting tool and workpiece or may not have a cutting tool in the same sense as traditional processes or utilize thermal or chemical energy for removing material from workpiece, are ultrasonic machining, water jet machining, electro discharge machining, laser machining and electrochemical machining just to name some.

Traditional machining operations and more specifically turning, milling, drilling and grinding will be discussed in the at hand book. These processes were selected as FEM bibliography pertaining to these processes is more extended than others. Although these machining processes exhibit a lot of differences which will be briefly presented below, a physical analysis including orthogonal and oblique cutting can be used to describe them.

Before moving on to modeling, a brief description of turning, milling and drilling is provided; grinding is discussed individually in Chap. 5. There are many variations of these processes but is out of the scope of this paragraph to describe all of them. Only some geometrical and kinematic features of the processes involved in modeling discussed later on needs to be clarified here. The interested reader is encouraged to search for more details on manufacturing processes in general and machining in particular and some excellent books are proposed [1, 3–9].

1.2.1 Turning

Turning is a process of removing excess material from the workpiece to produce an axisymmetric surface, using a single-point tool. In Fig. 1.1 the general case of cylindrical turning is depicted. The workpiece is rotated at a work speed provided by the machine tool (lathe), which is the cutting speed at which cutting is performed. In order to perform the material removal operation, the workpiece is advanced along the feed direction and a cylindrical surface is generated. Generally speaking, feed is defined as the displacement of the tool relative to the workpiece in the direction of feed motion per revolution or per stroke of the workpiece or

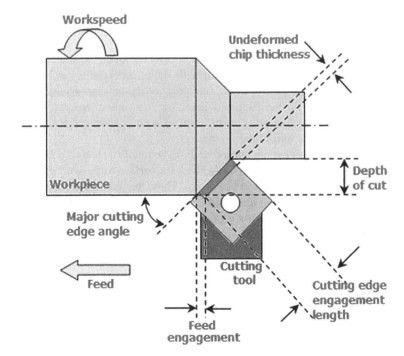

Fig. 1.1 Cylindrical turning

tool. In particular, in turning, feed is the distance that the tool advances in each workpiece revolution. For single-point cutting tools the feed is equal to the feed engagement, the instantaneous engagement of the workpiece and the cutting tool edge. The darker area on the cutting tool in Fig. 1.1, is the cross section of the uncut chip. The undeformed chip thickness in turning a_c, can be calculated through feed f and the major cutting edge angle κ_r, by equation:

$$a_c = f \sin \kappa_r \qquad (1.1)$$

The difference between the radius of the uncut workpiece and the machined workpiece is the depth of cut. The cutting tool contacts the workpiece along the cutting edge engagement length. It is obvious that for the case that the major cutting edge angle is 90°, the feed is equal to the undeformed chip thickness and the depth of cut is equal to cutting edge engagement length; the product of feed by depth of cut is the cross-sectional area of the uncut chip.

Turning cutting tools are also characterized by a rake and clearance angles at the tool edge. The rake angle is the inclination of the top face of the cutting edge to the surface being machined. The rake angle can be positive, usually for ductile materials, or negative, for high strength materials. The cutting edge also possesses side and front clearance angles to ensure that there is no contact of the major and minor flanks of the tool with the machined surfaces; if there was contact the surface integrity of the already machined surface would degrade. Finally, turning tools have a nose radius incorporated between the major and minor cutting edges. After milling and drilling are described some notes on the terminology and special characteristics considering cutting tools are given.

1.2.2 Milling

Milling has many different variants thus allowing for a wide variety of shapes to be processed, involving the machining of horizontal, vertical and inclined surfaces through horizontal or vertical milling machines. In Fig. 1.2, face milling, a process that can be utilized to machine profiles, pockets and slots, is depicted.

The workpiece is reduced in height by an amount equal to the axial depth of cut, over a width equal to the radial width of cut. The feed is the distance that the cutter advances across the workpiece per revolution. In milling the cutting tool has many cutting edges; in the case of Fig. 1.2, the cutting edges are four. Each cutting edge has a major cutting edge angle and creates a chip. For this multi-point tool, material removal is performed by the clockwise rotation of the tool. Cutting edge A located in point 1, in the specific time frame depicted in Fig. 1.2, enters the workpiece and starts creating the chip. This cutter will leave workpiece in point 2, having created a chip with an increasing thickness. Cutter B is already removing material while cutter D is the next in turn to continue this process. If the cutting tool, with cutters facing the opposite side, was rotating anticlockwise and thus entering at point 2 and leaving at point 1, the chip thickness would decrease with the edge's travel. For milling an average undeformed chip thickness can be determined. The difference between the minimum and maximum undeformed chip thickness depends on the cutting conditions.

1.2.3 Drilling

In this process a multi-point tool is used with the aim to create a hole. In Fig. 1.3, depicted is a tool with two flutes, therefore two cutting edges that is rotated and fed downwards along its axis of rotation. Drilling is a multi-point tool process, like milling, but each cutting edge is continuously engaged with the workpiece creating a chip, like turning. Each cutting edge is characterized by the major cutting edge angle of the tool. The tip of the drill is a chisel that advances inside the workpiece, pushing the material to be removed by the cutting edges.

The maximum cutting speed is reached in the outer radius of the major cutting edges and decreases to almost zero in the drill's center. The depth of cut of the process is the radius of the hole being drilled. The axial feed in drilling is expressed as feed per revolution.

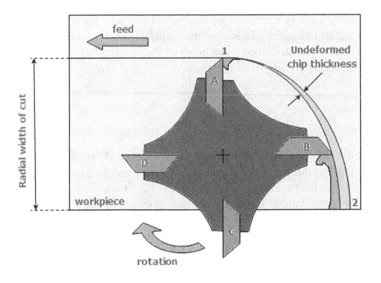

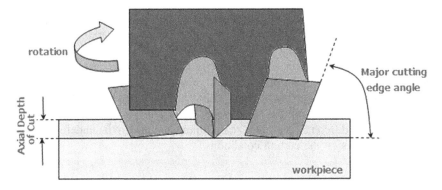

Fig. 1.2 Face milling

1.2.4 Discussion on Cutting Processes and Chip Formation

The brief description of turning, milling and drilling reveals the differences that they exhibit. However, it is more important for modeling to identify the similarities that allow for a unified approach. In Table 1.1, some chip formation terminology that is used in all three processes and the alternative term for each process is provided.

The terms described in Table 1.1 can be the same or different in every process but their overall effect is equivalent. An example is the feed which in turning and drilling is the distance travelled by the cutting edge per workpiece revolution, while in milling it refers to the distance travelled by the workpiece per cutting

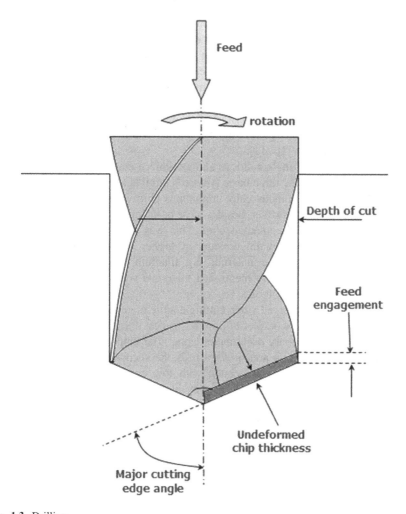

Fig. 1.3 Drilling

Table 1.1 Chip formation terminology for turning, milling and drilling [2]

Term	Turning	Milling	Drilling
Rake angle	Side rake angle	Radial rake angle	Rake angle
Major cutting edge angle	Approach angle	Entering angle	Point angle
Feed	Feed per revolution	Feed per edge[a]	Feed per revolution
Depth of cut	Depth of cut	Axial depth of cut	Hole radius

[a] In milling besides feed per edge, feed per minute or feed per tool revolution can be also used

edge. However, the relative movement between the cutting tool and the workpiece in order to create the chip is described in every case.

In turning, the case of the major cutting edge angle being 90° was described; it was concluded that feed and depth of cut are equal to the undeformed chip thickness and the cutting edge engagement length respectively. The former parameters refer to the machine tool movement while the latter to the chip and the way it is formed. It is usual to consider these terms that coincide in every case, although, this is not true, e.g. milling. For further reading on cutting tools, two books are recommended [10, 11].

Depending on the cutting conditions and workpiece material, different kinds of chip may be produced. The three basic types of chip that are typically encountered in metal cutting are continuous chip, continuous chip with built-up edge (BUE) and discontinuous chip. However, besides these basic chip types, another chip type is observed, namely the segmented chip. This was documented later than the previous three because it mainly occurs at higher cutting speeds and when machining difficult-to-machine materials, e.g. titanium alloys; machining at high speeds or machining the above mentioned materials was realized by the development of advanced cutting tools.

Continuous chip is observed when cutting ductile materials at high speeds. The chip is detached from the workpiece through shearing in front of the cutting tool in a zone referred to as primary deformation zone. The chip slides on the tool face and is deformed again; this area is known as the secondary deformation zone. Although continuous chip is associated with low cutting forces, long chip or curling of the chip are not desirable as they increase forces and worsen surface finish; in this case tools with chip breakers are used. In the second kind of chip, BUE is usually formed when ductile materials are machined at low cutting speeds. BUE is material piled up on the cutting edge, resembling the case of cutting with a blunt tool; BUE is associated with increased cutting forces and influences tool wear and surface finish. BUE may be removed with speed increase. Discontinuous chip usually occurs when machining brittle materials or ductile materials at very low speeds due to severe strain from the process. The chip breaks into small parts in the primary deformation zone, when the chip is only partly formed. The serrated or segmented chip, also referred as "shear localized", results in increased chip velocity, chip-tool friction and temperatures at the rake face of the tool that consequently provoke significant wear and tool life reduction. Although it is generally accepted that chip segmentation is energetically favorable, tool wear issues exist. However, tool life can be prolonged by optimizing cutting parameters, cutting conditions and machining strategy.

Most of the modeling work performed in machining pertains to continuous chip formation. It is considered the simplest type of chip formation, essentially under steady-state process conditions. However, models of the other chip types exist and will be discussed.

References

1. Trent EM, Wright PK (2000) Metal cutting. Butterworth-Heinemann, MA
2. Childs THC, Maekawa K, Obikawa T, Yamane Y (2000) Metal machining: theory and applications. Elsevier, MA
3. Grzesik W (2008) Advanced machining processes of metallic materials. Theory, modelling and applications. Elsevier, Oxford
4. Shaw MC (1984) Metal cutting principles. Oxford University Press, Oxford
5. Schey JA (1987) Introduction to manufacturing processes, 2nd edn. McGraw-Hill International Editions, New York
6. Kalpakjian S, Schmid SR (2003) Manufacturing processes for engineering materials, 4th edn. Prentice Hall, NJ
7. Groover MP (2002) Fundamentals of modern manufacturing. Materials, processes, and systems. Wiley, MA
8. Stephenson DA, Agapiou JS (2006) Metal cutting theory and practice, 2nd edn. CRC Press, FL
9. Boothroyd G, Knight WA (2006) Fundamentals of machining and machine tools, 3rd edn. CRC Press, FL
10. Smith GT (2008) Cutting tool technology: industrial handbook. Springer, London
11. Astakhov VP (2010) Geometry of single-point turning tools and drills. Fundamentals and practical applications. Springer, London

Chapter 2
Cutting Mechanics and Analytical Modeling

2.1 Questions and Answers on Machining Modeling

Prior to the description of the most important modeling methods and their features, it would be helpful to introduce some questions that may come to mind of those who want to use modeling, and attempt to give answers. Although some answers are already given in the previous chapter, a more elaborated approach is presented in this section. The questions raised apply to all kinds of modeling; the answers mostly concern FEM, without excluding all the other methods. In the next chapter some more questions and answers, this time solely for FEM, will be presented.

A first question would be: *what is modeling and what is simulation?* A model can be defined as an abstract system which is equivalent to the real system with respect to key properties and characteristics, and is used for investigations, calculations, explanation of demonstration purposes, which would otherwise be too expensive or not possible. A model permits general statements about elements, structure and behavior of a section of reality. Simulation is an imitation of a dynamic process in a model in order to obtain knowledge which can be transferred to reality. Both definitions, for model and simulation, are quoted from Ref. [1]; the former is from Brockhaus while the latter from VDI Guideline 3663.

An obvious question that may occur or has occurred to everybody reading this text would be: *Why model machining? What is the benefit coming out of this task?* Today, most of the researchers dealing with machining modeling perform it for its predictive ability. Important parameters of machining such as cutting forces, temperatures, chip morphology, strains and stresses can be calculated before actually any cutting is performed on a machine tool. The trial-and-error approach is far more laborious, costly and time-consuming. With modeling, resources are spared, optimization is achieved and cost is reduced. The above do not mean that experimental work is obsolete, since in most cases a validation of the model is needed and the only way to provide it is to actually test model results in real conditions and make comparisons. However, modeling reduces experimental work

A. P. Markopoulos, *Finite Element Method in Machining Processes*,
SpringerBriefs in Manufacturing and Surface Engineering,
DOI: 10.1007/978-1-4471-4330-7_2, © The Author(s) 2013

considerably. Furthermore, modeling and experiments add to the understanding of fundamental issues of machining theory. This forms a feedback loop vital for machining research since better understanding of the processes results in better models and so on. After all "understanding is the next best thing to the ability to predict" [2]. In a keynote paper by CIRP [3] two different "traditional schools" in machining modeling were identified, namely the one that treats modeling as an engineering necessity and another that treats modeling as a scientific challenge. On the long run both have to produce accurate models for the benefit of industry.

Who is, then, interested in machining modeling operations? The answer is the academia and the industry since there are benefits for both and the one depends on the other.

All these benefits are important and it seems that modeling is the solution for many problems. One may ask: *what are the drawbacks?* The answer is that it is the difficulties rather than the drawbacks that explain why modeling is not a panacea. The question could be rephrased to: *why is it difficult to model machining?* The answer lies in the fact that there are too many variables that need to be taken into account. First of all, there are a lot of machining operations and even though similarities do exist, many factors that are case sensitive make the proposal of a universal model not realistic. Even the orthogonal cutting system and shear plane models that are widely used are under criticism, as will be discussed in the next chapter. In the previous chapter a concise description of some machining operations was given in order to point out the similarities and differences that need to be accounted for in modeling, and that is only for traditional machining operations.

Secondly, difficulties arise from the fact that machining is still one of the least understood manufacturing operations. Machining typically involves very large stresses and strains in a small volume and at a high speed. The mechanisms of chip formation are quite complex, leading to equally complex theories and models that represent these theories. It is true that models always include simplifications in order to adequately embody theory but the danger of oversimplification is lurking. The result would be either inaccurate and thus erroneous results or models applicable for only very specific and confined cases. It should be notated that any kind of model is always applicable within the extremes of its input data. However, the area of application must be as wide as possible in order to have practical use. The mechanics of metal machining are briefly presented in the next paragraph. It can be observed that the application of the theory of plasticity on machining is far more complex than e.g. forming processes.

Finally, the variation of workpiece and tool properties and geometrical characteristics, machining conditions such as cutting speed, feed and depth of cut, the use of cutting fluids and the interaction of all the above in the same system increases the complexity of a model. The above are, generally speaking, the input data required to get a model started. Different input parameters will result in different output, significantly altering cutting forces, temperatures and chip morphology. In 1984 that "Metal Cutting Principles" by M. C. Shaw was published it seemed "next to impossible to predict metal cutting performance" [4], due to the complexity of model inputs and system interactions. This is where modern modeling techniques come to fill in the gap, as it will be pointed out in Chaps. 3 and 4 that are dedicated to FEM.

After the selection of the process, the properties and characteristics of the cutting tool and the workpiece and the determination of the cutting conditions, *what else is needed to build a model?* This answer depends on the selected kind of modeling. The finite element method for example would require meshing parameters to be determined, boundary conditions to be inserted and, depending on the formulation used, maybe a separation criterion for chip creation simulation, among others. It is obvious that this question cannot be answered unless a modeling technique is first selected.

Next question would be then: *what kind of modeling should one choose?* The answer would be the one that is able to provide a reliable answer to the variable/ output that it is looked for, with the available input data. There are five generic categories of modeling techniques available [1], i.e. empirical, analytical, mechanistic, numerical and artificial intelligence modeling. More complex models may require more input data; other models may not be able to predict a required parameter. Note, also, that the interest in predictive machining modeling has changed over the years due to advances in cutting technology. In the early days of metal cutting, tool wear was of utmost importance but nowadays the interest has shifted over to e.g. accuracy and determination of cutting forces, temperatures and the kind of produced chip. Furthermore, industry is interested in high speed machining and is environmentally conscious, requiring cutting fluids reduction or omission. Analytical models may predict output data, i.e. cutting forces, through equations requiring constants of workpiece material taken from databases, verified by experimental work, but the major drawback is that for out-of-the-ordinary cases no reliable results can be acquired. FEM on the other hand can perform coupled thermo-mechanical analysis but requires a considerable amount of computational power to produce accurate results. Artificial Intelligence techniques are usually simpler and faster models but provide results focused on a parameter or a specific area of the workpiece. It is true to say that the selection of the modeling technique depends heavily on information technology parameters, as well. Speaking for Finite Elements, more accurate representations of machining processes, e.g. 3D models, are coming true due to the fact that more powerful computers that can perform complicated calculation at an acceptable time are now available. Commercial software of FEM and especially for machining has qualified this technique to be the first choice for modeling machining operations, for many researchers.

Finally, *why would a model fail?* A model fails when it cannot predict accurately. However, it may also be considered not acceptable if it is not simple or fast enough for practical use. Many reasons may contribute to failure; lack of accurate input data, inadequate inclusion of all important parameters and misuse of a modeling technique are the most common reasons for that, as will be exhibited in the next chapter.

Machining technology cannot rely on the craftsmanship of technicians or time-consuming experiments in order to advance and meet the requirements of modern industry. Nowadays, machining is more science than art. A scientific approach is required and modeling offers solutions. Modern modeling techniques, such as FEM, in close cooperation with computer advances are able to provide reliable results in a timely manner, justifying the many publications and research groups that are dealing with them.

2.2 Orthogonal and Oblique Cutting

The chip flow in all wedged-tool machining processes can be described, in theory, in a common way by two different cutting schemes termed orthogonal cutting and oblique cutting, depicted in Figs. 2.1 and 2.2 respectively. In orthogonal cutting the cutting edge of the tool is perpendicular to the direction of relative workpiece-cutting tool motion and also to the side face of the workpiece. From the relative movement of workpiece and cutting tool, a layer of material in the form of chip is removed. In order to continue removing material at a second stage, the tool is taken back to its starting position and fed downwards by the amount f, the feed of the process. Perpendicular to f, d is the depth of cut, which is smaller than or equal to the width of the tool edge. The surface along which the chip flows is the rake face of the tool. The angle between the rake face and a line perpendicular to the machined surface is called rake angle γ. The face of the tool that is near the machined surface of the workpiece is the flank face. The angle between the flank face of the tool and the workpiece is called clearance angle α. The angle between the rake face and the flank face is the wedge angle β. The sum of the three angles is always equal to 90°, thus:

$$\alpha + \beta + \gamma = 90° \tag{2.1}$$

In Fig. 2.1 a positive rake angle is shown; in the same figure the direction for a positive or a negative rake angle is shown. For negative rake angles, the tools possess a wider wedge angle. As pointed out in Sect. 1.2.1, a positive rake angle is used for ductile materials since a "weaker" tool, with smaller wedge angle, will suffice to perform the cutting operation. For high-strength materials, rake angle is chosen to be negative, thereby increasing the wedge angle and creating a stronger cutting edge. However, stronger cutting edge has the disadvantage of requiring greater power consumption and needing a robust tool-workpiece set-up to compensate for the vibrations. The flank face of the tool does not participate in chip removal; it ensures that the tool does not rub on the newly machined surface and affects its quality. However, the clearance angle affects the cutting tool wear rate. If the tool's clearance is too large it will weaken the wedge angle of the tool, whereas if too small, it will tend to rub on the machined surface.

Orthogonal cutting represents a two-dimensional mechanical problem with no side curling of the chip considered. It represents only a small fragment of machining processes, i.e. planning or end turning of a thin-walled tube. However, it is widely used in theoretical and experimental work due to its simplicity. Because of its 2D

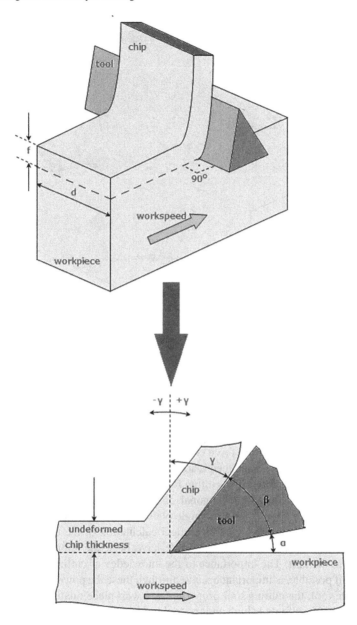

Fig. 2.1 Orthogonal cutting

nature many independent variables are eliminated, e.g. two cutting forces are only identified to orthogonal cutting problems. On the other hand, oblique cutting, where the cutting tool is inclined by angle λ, as it can be seen in Fig. 2.2, corresponds to a three-dimensional problem with more realistic chip flow representation but more

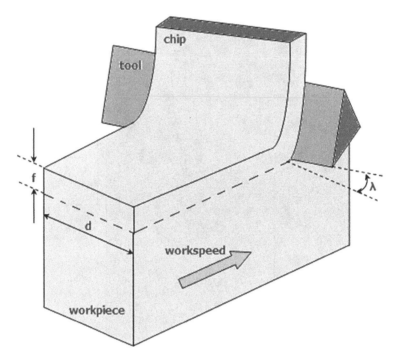

Fig. 2.2 Oblique cutting

complex analysis, i.e. three force components are present and chip curling is accounted for.

In oblique cutting that is a more general case than orthogonal, there are three mutually perpendicular cutting force components. If a coordinate system based on the directions of work speed and feed is adopted, the cutting force, the feed force and the back force are considered. The cutting force is usually the largest and back force the smallest component. For orthogonal cutting, the third force component is ignored, so the force system lies in a single plain, normal to the cutting edge of the tool. The measurement and/or the theoretical calculation of the two cutting force components as well as their resultant force have been the subject of numerous researches in the past. The importance of the knowledge of cutting forces, prior to machining if possible, is important because through these the power requirements of the machine tool, the cutting tool properties and workpiece quality are estimated. For example, if feed force is high and the tool holder is not stiff enough, the cutting edge will be pushed away from the workpiece surface, causing lack of dimensional accuracy. Furthermore, determination of cutting forces can easily lead to the calculation of other parameters, e.g. stresses.

There are two deformation areas distinguished in machining, namely the primary and the secondary deformation zones, see Fig. 2.3; the deformation zones thickness, chip thickness and shear angle are not depicted in any scale in this figure, only the locations are roughly indicated. The primary deformation zone is included in the

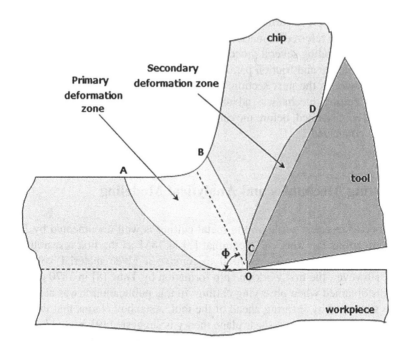

Fig. 2.3 Primary and secondary deformation zone and shear plane angle

OAB area. The workpiece material crossing the OA border undergoes large deformation at high strain rates and exits the zone at OB border, work hardened. It is determined by microscopic examination and experiments that chips are produced by shear within this region. Most of the experimental studies conclude that this zone is of average thickness of about one tenth of chip thickness [4]. The secondary deformation zone is included in OCD. Along OD, the contact length between the rake face of the tool and the chip, the material is deformed due to intensive interfacial friction. The secondary deformation zone is characterized by two regions, the sticking region, closer to the cutting tool tip and the sliding region, above the previous one [5]. In the sticking region, material adheres to the tool and as a result shear within the chip is observed. Both deformation zones are characterized by temperature rise due to severe plastic deformation in the primary and due to friction in the secondary deformation zone. Furthermore, high cutting speeds do not allow for heat conduction to take place and heat is concentrated at a small area around the cutting tool edge. Strain hardening due to deformation and softening due to temperature alter the chip formation characteristics in every step of its formation. The friction coefficient is very hard to be measured in the secondary deformation zone. Several theories are proposed for the calculation of friction, discussed in another part of this book.

A simplified approach proposes that shearing in the primary deformation zone takes place along a shear plane, characterized by shear angle ϕ, between the shear

plane and the workpiece surface. Although this single shear plane model is criticized, it is usually referred in machining handbooks due to its simplicity and it is
the basis for calculating several process parameters. In any case, it is imperative to
estimate shear angle and friction parameters in order to calculate cutting forces, as
explained above. In the next section, an overview of the theoretical approach of
machining, cutting mechanics, advances in cutting mechanics and analytical
models will be discussed, before moving on to FEM analysis, since all these topics
are closely connected.

2.3 Cutting Mechanics and Analytical Modeling

The history of research pertaining to metal cutting is well documented by Finnie
[6] who pinpoints the work of Cocquilhat [7] in 1851 as the first research in the
area of measuring the work required to remove a given material volume by
drilling. However, the first work on chip formation by Time [8] in 1870 presented
the results obtained when observing cutting. In this publication it was argued that
the chip is created by shearing ahead of the tool. Astakhov claims that this is one
of the first publication that a shear plane theory is suggested [9], probably the first
being the one by Usachev in 1883 [10]. It is also shown that there is no contradiction between Time and Tresca [11]; Tresca argued that the chip in metal cutting
is produced by compression ahead of the tool. Zvorykin [12] was the first to
provide physical explanation for this model; his work resulted to an equation
predicting the shear angle. In 1881, Mallock [13] also identified the shearing
mechanism in chip formation and emphasized the importance of friction in the
tool-chip interface. However, it was the work of Ernst and Merchant [14] in 1941
that made the shear plane model popular; most of the fundamental works on metal
cutting mechanics reference this paper and many analytical models of orthogonal
cutting still use the relations derived from this work. In the following paragraphs
some key points of analytical modeling and advances in mechanics of cutting will
be discussed.

Analytical models, only briefly described, are considered the predecessors of
numerical models. This is by no way meant to say that numerical models
substituted analytical modeling, since a lot of researchers still are working on this
subject and the value of these models is paramount. It is meant to say that they
have the same origins and form the basis on which FEM models and simulations
are made. In another paragraph it will be discussed what the benefits, and some
drawbacks as well, are from choosing numerical modeling over analytical. As it
can be concluded analytical models are quite controversial and up to date there is
no model universally accepted or employed. The subject cannot be portrayed in its
full length within this book. However, many excellent books on mechanics of
machining can be found and it is the author's opinion that they should be
considered by the prospective modeler before moving on to numerical or any other
kind of machining modeling [5, 9, 15–18]. These books include theory of

plasticity, slip-line theory, shear zone models and usually chapters on numerical modeling as well, among other subjects.

2.3.1 Lower and Upper Bound Solutions

Most of the analytical modeling works aim at producing equations that can determine cutting forces, without any experimental work; that is useful since other parameters can be derived by cutting forces and analysis in tool wear, surface integrity and workpiece quality can be carried out. The problem involved in the determination of the cutting forces, when the cutting conditions are known, ends up in determining a suitable relationship between the shear angle, the rake angle and the friction coefficient. Several methods have been employed that either overestimate or underestimate the results; the real value of the cutting forces probably lies between these lower and upper bounds.

Lower bound solutions employ the principle of maximum work, i.e. the deformation caused by the applied stresses results to maximum dissipation of energy. The system tends to reach the state of minimum energy compatible with the equilibrium and yield conditions. Any other statically compatible system will produce work that is either equal or less than that of the actual system.

In the upper bound solutions the strain increments of a fully plastic body rather than the stress equilibrium is considered. The principle of maximum work is employed in this case from the point of view of strain. The material is incompressible, thus the plastic volume remains constant. An element of this system deforms so that it exhibits maximum resistance. If the stresses are deduced from deformations imposed by the kinematic conditions, the estimation of their values will be equal or greater than the ones actually occurring.

2.3.2 Shear Plane Models

Shear plane models are closely connected to the theory of Ernst and Merchant, as mentioned above. This shear model was based on the so-called card model of Piispanen [19, 20]. The chip is formed by shear along a single plane inclined at an angle ϕ. The chip is straight and has infinite contact length with the tool. The shear stress along the shear plane is equal to the material flow stress in shear.

The chip is assumed to be a rigid body in equilibrium. The equilibrium refers to the forces on the chip-tool interface and across the shear plane. In Fig. 2.4 the Merchant's circle force diagram is given. All forces are shown acting at the tool tip.

The resultant force F is resolved in components F_N and F_F that are normal to the tool face the former and along the tool face the latter. It is also resolved to F_{SN} and F_S that are normal to and along the shear plane respectively. Finally, it can also be resolved into components F_c, the cutting force, and F_t the feed or thrust force. Furthermore, the rake angle γ, the shear angle ϕ and the mean angle of friction

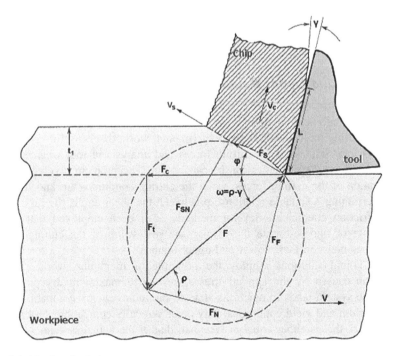

Fig. 2.4 Merchant's circle

between chip and tool ρ are shown. The friction angle ρ is related to the friction coefficient μ through equation:

$$\rho = \arctan(\mu) = \arctan(F_F/F_N) \tag{2.2}$$

According to Ernst and Merchant's theory, an upper bound one, a shear angle needs to be found that the cutting work will reduce to a minimum. In other words, since the work is proportional to the cutting force F_c, an expression of the cutting force with the shear angle needs to be found and then obtain the ϕ for which F_c is a minimum. From Fig. 2.4, it can easily be concluded that:

$$F_S = F \cos(\phi + \rho - \gamma) \tag{2.3}$$

Furthermore, the same force component can be calculated in relation to the shear strength of the workpiece material on the shear plane τ_S, the cross-sectional area of the shear plane A_S and the cross-sectional area of the undeformed chip A_C, via the following equation:

$$F_S = \tau_S A_S = \frac{\tau_S A_C}{\sin \phi} \tag{2.4}$$

Thus from Eqs. 2.3 and 2.4 it is:

$$F = \frac{\tau_S A_C}{\sin \phi} \cdot \frac{1}{\cos(\phi + \rho - \gamma)} \tag{2.5}$$

Geometrically it is deducted that:

$$F_c = F \cos(\rho - \gamma) \tag{2.6}$$

Combining Eqs. 2.5 and 2.6 it may be concluded that:

$$F_c = \frac{\tau_S A_C}{\sin \phi} \cdot \frac{\cos(\rho - \gamma)}{\cos(\phi + \rho - \gamma)} \tag{2.7}$$

If the last equation is differentiated with respect to ϕ and equated to zero, it is possible to calculate a shear angle for which the cutting force is minimum. The equation is:

$$2\phi + \rho - \gamma = \pi/2 \tag{2.8}$$

This equation agreed poorly with experimental results of metal machining. Merchant attempted an alternative solution [21]. When Eq. 2.7 was differentiated it was assumed that A_c, γ and τ_S where independent of ϕ. In the new theory, deformation and friction are reflected through a change of the force acting in the direction perpendicular to the plane of shear, thus the normal stress σ_S of the shear plane affects the shear stress τ_S. In the modified analysis a new relation is included:

$$\tau_S = \tau_o + k\sigma_S \tag{2.9}$$

This relation is known as the Bridgman relation and k is the slope of the $\tau - \sigma$ relation; the shear stress increases linearly with an increase in normal strength and the lines intersects the shear stress axis at τ_o. With this revised theory the new result for shear angle is:

$$2\phi + \rho - \gamma = C \tag{2.10}$$

C is a constant that depends on the workpiece material.

2.3.3 Slip-Line Field Models

Stress analysis in a plane strain loaded material indicates that at any point there are two orthogonal directions that the shear stresses are reaching a maximum, but these directions can vary from point to point. A line, which generally speaking is curved, tangential along its length to the maximum shear stress is called a slip-line; a complete set of slip-lines in a plastic region forms a slip-line field. The slip-line field theory must follow rules that allow the construction of a slip-line field for a particular case. First of all, the boundary between a part of a material that is plastically loaded and another that has not yielded is a slip-line. In machining, the

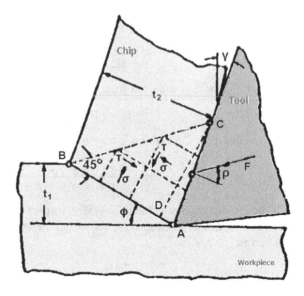

Fig. 2.5 Lee and Shaffer's
slip-line field theory for
orthogonal cutting

borders of the primary deformation zone with the workpiece on the one side and
the chip on the other are slip-lines. Similarly, a slip-line is the border between the
secondary deformation zone and the chip. Another rule is that slip-lines must
intersect free surfaces at 45° angle.

Lee and Shaffer's work was the first contribution of the slip-line field models of
chip formation [22]. It was the result of applying simplified plasticity analysis to
metal cutting, more specifically to orthogonal cutting with continuous chip. It was
assumed that in this plane strain conditions, the workpiece material is rigid perfectly
plastic, i.e. the elastic strain is neglected during deformation and once the yielding
point is exceeded deformation takes place at constant stress for varying strains, strain
rates and temperatures. The constructed slip-line field is shown in Fig. 2.5.

In this lower bound solution all deformations take place in a stress field bounded
by rigid bodies; this stress field transmits the cutting forces from the shear plane to
the chip resulting in the triangular plastic zone ABC. In this region no deformation
occurs but the material is stressed to its yield point, so that the maximum shear stress
is the shear stress on the shear plane. The two directions of the maximum shear stress
are indicated by the slip-lines. The shear plane AB is the one set of slip-lines because
the maximum shear stress must occur along the shear plane. Furthermore, BC can be
regarded a free surface since no forces act on the chip after BC, stresses cannot be
transmitted from there. Thus, according to the second rule mentioned above, ABC is
equal to $\pi/4$. Assuming that stresses act uniformly at the chip-tool interface, normal
stresses will meet the boundary at angles ρ and $\rho + \pi/2$. Maximum shear stresses
are $\pi/4$ to the direction of normal stresses and thus ACB is $(\pi/4) - \rho$. The shear
angle can be calculated by equation:

$$\phi + \rho - \gamma = \pi/4 \tag{2.11}$$

It is evident that when the mean angle of friction between chip and tool is $\pi/4$ and the rake angle is zero, shear plane angle is also zero, which is not possible. Lee and Shaffer proposed a solution for this case of high friction and low rake angle, assuming built-up edge formation.

The slip-line theory was also used by other researchers who suggested curved AB and CD boundaries [23, 24]. These models reveal the non-uniqueness of machining processes; different chip shapes and thicknesses result from the same specified conditions. The non-uniqueness of the possible solutions is a significant limitation, resulting mainly by the rigid plastic workpiece material assumption.

At this point it would be interesting to make a note on the work of Zorev in relation to the slip-line field theory [5]. Zorev proposed an approximate form of the shear lines in the plastic zone as it can be seen in Fig. 2.6 on top. This is a qualitative model for which no solution is provided. However, a simplified form was proposed as shown in the same figure; in this simplified model the curved shear lines are replaced by straight ones and it is assumed that no shearing occurs along the shear lines adjacent to the tool rake face. By using geometrical relationships a generalized solution is derived as:

$$2\phi_{sp} + \rho - \gamma \approx (\pi/2) - \psi_{sp} \qquad (2.12)$$

In this equation the ϕ_{sp}, the specific shear angle is introduced and ψ_{sp} is the angle of inclination of the tangent to the outer boundary of the plastic zone. The interesting about this solution is that if various values of ψ_{sp} are substituted, the shear angle relations by other researchers are derived, i.e. for ψ_{sp} equal to zero, representing the single shear plane model, the Ernst and Merchant solution is obtained, for $\psi_{sp} = C_1$ and $C = (\pi/2) - C_1$ the modified Merchant solution is obtained and for $\psi_{sp} = \rho - \gamma$ the Lee and Shaffer solution is derived.

2.3.4 Shear Zone Models

The next step in analytical modeling was to enhance some features that were neglected or simplified in previous models but play an important role in metal cutting. Most shear plane models assume that shear stress on the shear plane is uniform, no strain hardening is considered and that friction along the cutting tool-chip interface is characterized by a constant friction coefficient; this last assumption is in contradiction with experimental data. If it is assumed that deformation takes place in a narrow band centered on the shear plane, more general material assumptions can be used. The effects of yield stress varying with strain and sometimes with strain rate and temperature were considered and simplification of the equilibrium and flow was achieved. Pioneering work in this area is associated with the work of Oxley. Based on experimental data, where the plastic flow patterns are observed, it is assumed that the shear zone thickness is about one tenth of the shear zone length. Then strain rate and strain at every point in the primary deformation zone can be calculated; strain rates are derived from

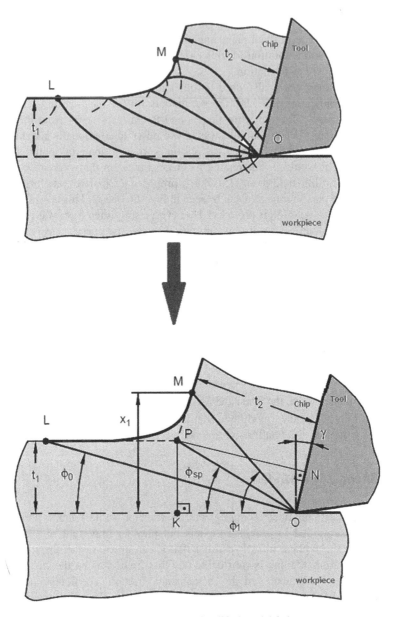

Fig. 2.6 Zorev's qualitative model on *top* and simplified model *below*

variations in the velocity with respect to the position and strains are calculated by integrating strain rates with respect to time along the streamlines of the flow. Similar assumptions are used to compute strain rates and strains in the secondary deformation zone.

Table 2.1 Shear angle formulas

Model	Formula	Year
Ernst–Merchant	$\phi = \frac{\pi}{4} - \frac{1}{2}(\rho - \gamma)$	1941
Merchant	$\phi = \frac{C}{2} - \frac{1}{2}(\rho - \gamma)$	1945
Stabler	$\phi = \frac{\pi}{4} - \rho + \frac{\gamma}{2}$	1951
Lee–Shaffer	$\phi = \frac{\pi}{4} - (\rho - \gamma)$	1951
Hucks	$\phi = \frac{\pi}{4} - \frac{a\tan(2\mu)}{2} + \gamma$	1951
Shaw et al.	$\phi = \frac{\pi}{4} - (\rho - \gamma) \pm \eta$	1953
Sata	$\phi = \frac{\pi}{4} - \gamma \pm \frac{\gamma - 15°}{2}$	1954
Weisz	$\phi = 54.7° - (\rho - \gamma)$	1957
Kronenberg	$\phi = a\cot\left[\dfrac{e^{\mu\left(\frac{\pi}{2}-\gamma\right)} - \sin\gamma}{\cos\gamma}\right]$	1957
Colding	$\phi = a\tan\left[-\dfrac{2\left(\frac{F}{h}+2\right)}{\left(\frac{F}{h}+1\right)}\cot(2\Omega) - (\rho - \gamma)\right]$	1958
Oxley	$\phi = a\tan\left[1 + \frac{\pi}{2} - 2\phi + \frac{\cos 2(\phi-\gamma)}{\tan\rho} - \sin 2(\phi - \gamma)\right] - (\rho - \gamma)$	1961
Sata–Yoshikawa	$\phi = a\cot\left\lfloor\cot\theta + \frac{\cos\theta}{\sin(\theta+\gamma)}kL\right\rfloor$	1963
Das–Tobias	$D = \frac{\cos(\rho-\gamma)}{\cos(\rho-\gamma+\phi)}$	1964

The shear zone models are an obvious improvement over the preceding models. Many additions to the first model proposed by Oxley have been reported. A full account of these developments would be out of the scope of this work; a detailed description of Oxley's works is given in [17].

2.3.5 Discussion on Analytical Modeling of Machining

The analysis presented here is not at all a complete one; with more than 50 shear angle solutions identified in the relevant literature as it is reported in [3] this would be impossible within this book. However, an outline of the most important models and the development over the years is presented. Furthermore, in Table 2.1 some shear angle formulas are gathered. In the following lines some drawbacks in the analytical modeling procedure are discussed.

The single shear plane model has been criticized over the years and experimental data do not correlate with the theory results. Astakhov [25] summarized the major inherent drawbacks of the single shear plane model as being the infinite strain rate, the unrealistic high shear strain that is in contradiction with material testing results, the rigid perfectly plastic workpiece material assumption, the improper accounting for the resistance of the processed workpiece material, the perfectly sharp cutting edge of the tool and the fact that there is no contact on the tool flank surface that are not realistic for common practice and the inapplicability of the model in brittle material machining. Furthermore, for the Ernst and Merchant theory, drawbacks include the incorrect velocity and force diagrams presented and the assumption of constant friction coefficient. However, this model is still in use by researchers due to its simplicity.

Slip-line solutions like the ones presented in Sect. 2.3.3 also have poor correlation with experimental results and no strain hardening is considered. Furthermore, the non-uniqueness of the models raises criticism on the results. Finally, Zorev's general model is based on geometrical considerations and no principle of mechanics of materials or physical laws are used. It is argued that all solutions related to this model, including Ernst and Merchant and Lee and Shaffer theory have little to do with physics and the mechanics of metal cutting [25].

The analyses already presented pertained only to orthogonal cutting with continuous chip. However, the shear plane model has been extended to three dimensions [26] and the slip-line model has been proposed for oblique cutting [27]. A three-dimensional analysis similar to the work of Oxley has been presented by Usui [28–30], which includes secondary cutting edge and nose radius effects; the results apply to turning, milling and groove cutting. However, both Oxley's and Usui's models are quite complex and for their application stress and strain data at the strain rates and temperatures encountered in metal machining are needed. The lack of these data is a significant drawback. These are the reasons that these models, although more complete than all the others since they include temperature effects and can be used in tool wear and segmented chip formation modeling and are in agreement with experimental data, are not widely used outside the research groups that they developed them. Nevertheless, Usui's tool wear estimation algorithm is integrated into finite element models for the prediction of tool wear; the commercial FEM software Third Wave AdvantEdge has the option of using this algorithm in the analyses it can perform.

Finally, another form of modeling for cutting force models will be briefly discussed here, namely Mechanistic modeling; a review can be found in [31]. This kind of modeling is not purely analytical because it is based on metal cutting mechanics but also depends on empirical cutting data; it is a combination of analytical and experimental modeling techniques. Such an approach avoids the complications of incorporating parameters such as shear angle and friction angle, by using experimental force data and it is suitable for use in oblique cutting and various cutting processes.

References

1. Grzesik W (2008) Advanced machining processes of metallic materials. Theory modelling and applications. Elsevier, Oxford
2. Shaw MC (1984) Metal cutting principles. Oxford University Press, Oxford
3. van Luttervelt CA, Childs THC, Jawahir IS, Klocke F, Venuvinod PK (1998) Present situation and future trends in modeling of machining operations. In: Progress report of the CIRP working group "modeling and machining operations". Annals of the CIRP, vol 47/2, pp. 587–626
4. Stephenson DA, Agapiou JS (2006) Metal cutting theory and practice, 2nd edn. CRC Press, FL

5. Zorev NN (1966) Metal cutting mechanics. Pergamon Press, Oxford
6. Finnie I (1956) Review of the metal cutting analyses of the past hundred years. Mech Eng 78(8):715–721
7. Cocquilhat M (1851) Expériences sur la Resistance Utile Produite dans le Forage. Annales Travaux Publics en Belgique 10:199–215
8. Time I (1870) Resistance of metals and wood to cutting (in Russian). Dermacow Press House, St. Petersbourg
9. Astakhov VP (1999) Metal cutting mechanics. CRC Press, FL
10. Usachev YG (1915) Phenomena occurring during the cutting of metals—review of the studies completed (in Russian). Izv. Petrogradskogo Politechnicheskogo Inst., XXIII(1)
11. Tresca H (1873) Mémoires sur le Rabotage des Métaux. Bulletin de la Société d' Encouragement pour l' Industrie Nationale, 585–607
12. Zvorykin KA (1896) On the force and energy necessary to separate the chip from the workpiece (in Russian). Vetnik Promyslennostie, 123
13. Mallock A (1881–1882) The action of cutting tools. Proc Roy Soc Lond 33:127–139
14. Ernst H, Merchant ME (1941) Chip formation, friction and high quality machined surfaces. Surface treatment of metals. Am Soc Met 29:299–378
15. Childs THC, Maekawa K, Obikawa T, Yamane Y (2000) Metal machining: theory and applications. Elsevier, MA
16. Bhattacharyya A (1994) Metal cutting theory and practice. New Central Book Agency Ltd, Kolkata
17. Oxley PLB (1989) The mechanics of machining: an analytical approach to assessing machinability. Ellis Horwood, Chichester
18. Dixit PM, Dixit US (2008) Modeling of metal forming and machining processes by finite element and soft computing methods. Springer, London
19. Piispanen V (1937) Lastunmuodostumisen Teoriaa. Teknillinen Aikakauslehti 27:315–322
20. Piispanen V (1948) Theory of formation of metal chips. J Appl Phys 19:876–881
21. Merchant ME (1945) Mechanics of the metal cutting process II. Plasticity conditions in orthogonal cutting. J Appl Phys 16(6):318–324
22. Lee EH, Shaffer BW (1951) The theory of plasticity applied to a problem of machining. Trans ASME—J Appl Mech 18:405–413
23. Kudo H (1965) Some new slip-line solutions for two-dimensional steady-state machining. Int J Mech Sci 7:43–55
24. Dewhurst P (1978) On the non-uniqueness of the machining process. Proc Roy Soc Lond A360:587–610
25. Astakhov VP (2005) On the inadequacy of the single-shear plane model of chip formation. Int J Mech Sci 47:1649–1672
26. Shaw MC, Cook NH, Smith PA (1952) The mechanics of three-dimensional cutting operations. Trans ASME 74:1055–1064
27. Morcos WA (1980) A slip line field solution of the free continuous cutting problem in conditions of light friction at chip-tool interface. Trans ASME—J Eng Ind 102:310–314
28. Usui E, Hirota A, Masuko M (1978) Analytical prediction of three dimensional cutting process. Part 1. Basic cutting model and energy approach. Trans ASME—J Eng Ind 100:222–228
29. Usui E, Hirota A (1978) Analytical prediction of three dimensional cutting process. Part 2. Chip formation and cutting force with conventional single-point tool. Trans ASME—J Eng Ind 100:229–235
30. Usui E, Shirakashi T, Kitagawa T (1978) Analytical prediction of three dimensional cutting process. Part 3. Cutting temperature and crater wear of carbide tool. Trans ASME—J Eng Ind 100:236–243
31. Ehmann KF, Kapoor SG, DeVor RE, Lazoglu I (1997) Machining process modeling: a review. Trans ASME—J Manuf Sci Eng 119:655–663

Chapter 3
Finite Element Modeling

3.1 Questions and Answers on Finite Element Modeling

In this chapter, general concepts of FEM are presented. Some advantages and disadvantage of the method are discussed, the various methods available are analyzed, a bibliographical review is presented and FEM programs are discussed. Following the questions and answers of Chap. 2, another similar discussion is made here, with the difference that only finite element modeling is concerned.

Previous chapters have pointed out the difficulties that are associated with modeling machining processes. First of all, the strain rates observed are very high; this holds true for even low cutting speeds. Furthermore, the plastic deformation takes place in small regions, the primary and secondary deformation zones, around the cutting edge, making difficult the selection of the appropriate boundary conditions. There is not a unified and generally accepted theory pertaining to the exact chip formation mechanism, mainly due to the phenomena taking place in the deformed regions. In many analytical models that are proposed, the strain hardening of the workpiece material is not included in the analysis, although it plays a significant role, as is concluded from experimental results. Additionally, the temperature rise in the region due to plastic deformation and friction induce material softening and alter the workpiece material properties in relation to strain rates and temperatures. Data for the workpiece material for varying temperature and strain rate at the levels which occur in metal machining are not easily found in the literature. On top of this, non-linear situation, the temperature rise needs to be taken into account to the various calculations performed, which means that besides the mechanical problem, a heat transfer problem must be dealt with simultaneously. This calls for a method that is able to perform coupled analysis. So, *what kind of modeling is the most appropriate for overcoming these problems?*

The finite element method appears to be the most suitable method for this task. Due to its inherent characteristics it can solve non-linear problems and with advances in computers and the use of commercial software it can readily perform

A. P. Markopoulos, *Finite Element Method in Machining Processes*,
SpringerBriefs in Manufacturing and Surface Engineering,
DOI: 10.1007/978-1-4471-4330-7_3, © The Author(s) 2013

coupled thermo-mechanical analysis. This kind of numerical modeling has already been used with success in many scientific and technological areas, modeling of manufacturing processes being one of them [1–3]. Still, chip formation is difficult to be modeled. Except the physical phenomena explained above two more challenges need to be addressed. The first one is to provide accurate data to the model; this is common sense however, it can be problematic. The second is to actually choose a finite elements method, meaning that there are different approaches or strategies proposed for metal machining modeling with FEM pertaining to formulation, treatment of friction, material behavior, iteration scheme etc. used for approximating a solution; the combinations that have already been tried by researchers are numerous. These approaches are discussed in the following paragraphs.

What is FEM? The Finite Element Method is the most used numerical technique employed in metal machining, other numerical methods being the Finite Differences Method (FDM) and Boundary Elements Method (BEM). In the finite element method the basic principle is the replacement of a continuum by finite elements forming a mesh; this procedure is called discretization. Each finite element is simpler in geometry and therefore easier to analyze that the actual structure. Every finite element possesses nodes where the problem initial and boundary conditions are applied and the degrees of freedom are calculated; the finite elements are connected to one another in nodes. Between the nodes, problem variables are derived by interpolation. The problem variables as well as properties applied on the nodes of each element are assembled and global relations are formatted. Usually, the analysis involves a great number of algebraic equations to determine nodal degrees of freedom and that is why a personal computer is employed for processing.

The discretization may be performed by many element types, of various shapes that have many nodes, i.e. triangular with 3 nodes or quadrilateral with 4 or 8 nodes see Fig. 3.1. Different kinds of elements may be combined in the analysis, given that certain rules are followed, i.e. adjacent elements share nodes and no nodes are unused. Function φ may represent various physical quantities, e.g. displacements for stress analysis, voltage for electric field or temperature for heat conduction. A polynomial function represents φ; a linear polynomial $\varphi = \alpha_1 + \alpha_2 x + \alpha_3 y$ for triangular element with three nodes, the bilinear function $\varphi = \alpha_1 + \alpha_2 x + \alpha_3 y + \alpha_4 xy$ for the quadrilateral element with 4 nodes, α_i, $i = 1-4$, being appropriate constants for the above interpolations. The $\varphi = \varphi(x,y)$ at each node of the element can be calculated.

How can the modeler decide what kind of discretization to do? The parameters to be decided are the kind of elements, how they are placed within the given geometry and the number of elements to be used in the analysis. The answer is somehow complicated. An element behaves well in a problem and may return poor results in another. The elements must be suitable for the physical problem that they face in the analysis and be suitable to cover the given geometry. Several meshing algorithms exist for that. The more finite elements used, i.e. finer the mesh is, the more accurate the model can be but with an increase in computational effort and

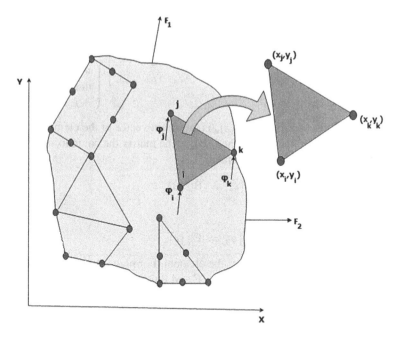

Fig. 3.1 Typical finite elements

consequently to analysis time. A modeler needs experience to determine these parameters. In the next paragraphs some more discussion will be given on how the mesh is applied in FEM machining models and some advanced procedures used for mesh rearrangement and refinement.

How can FEM calculate strains and stresses? A brief overview of FEM and its concepts is presented. For more information on FEM a book dedicated to this method is proposed, e.g. [4, 5]. The case of a thin plate loaded in its plane by external forces will be considered, see Fig. 3.1. The displacements of the triangular element shown in the same figure can be expressed by the linear interpolations:

$$u_x = \alpha_1 + \alpha_2 x + \alpha_3 y \text{ and } u_y = \alpha_4 + \alpha_5 x + \alpha_6 y \qquad (3.1)$$

Strain can be calculated by nodal displacements as:

$$\varepsilon_{xx} = \frac{\partial u_x}{\partial x}, \ \varepsilon_{yy} = \frac{\partial u_y}{\partial y} \text{ and } \gamma_{xy} = \frac{\partial u_x}{\partial y} + \frac{\partial u_y}{\partial x} \qquad (3.2)$$

Or in a matrix operator form:

$$
\left\{ \begin{array}{c} \varepsilon_{xx} \\ \varepsilon_{yy} \\ \gamma_{xy} \end{array} \right\} = \left[\begin{array}{cccccc} \partial/\partial x & 0 & \partial/\partial x & 0 & \partial/\partial x & 0 \\ 0 & \partial/\partial y & 0 & \partial/\partial y & 0 & \partial/\partial y \\ \partial/\partial y & \partial/\partial x & \partial/\partial y & \partial/\partial x & \partial/\partial y & \partial/\partial x \end{array} \right] \left\{ \begin{array}{c} u_{x,i} \\ u_{y,i} \\ u_{x,j} \\ u_{y,j} \\ u_{x,k} \\ u_{y,k} \end{array} \right\} \tag{3.3}
$$

In a more compact form, where $\{\varepsilon\}_e$ is the strain vector of the element, $\{u\}_e$ the displacement vector of the element and $[B]_e$ is the matrix the contents of which are shown in Eq. 3.3, for the same element, it is:

$$
\{\varepsilon\}_e = [B]_e \{u\}_e \tag{3.4}
$$

Hooke's law is expressed as:

$$
\{\sigma\}_e = [E]_e \{\varepsilon\}_e \tag{3.5}
$$

Where $\{\sigma\}_e$ is the stress vector of the element. For plane stress conditions of an isotropic material with Young's modulus E and Poisson's ratio v, it can be written:

$$
\left\{ \begin{array}{c} \sigma_{xx} \\ \sigma_{yy} \\ \tau_{xy} \end{array} \right\} = \frac{E}{1-v^2} \left[\begin{array}{ccc} 1 & v & 0 \\ v & 1 & 0 \\ 0 & 0 & \frac{1-v}{2} \end{array} \right] \left\{ \begin{array}{c} \varepsilon_{xx} \\ \varepsilon_{yy} \\ \gamma_{xy} \end{array} \right\} \tag{3.6}
$$

Finally, forces can be calculated as:

$$
\{F\}_e = t\Delta [B]_e^T [E]_e [B]_e \{u\}_e \tag{3.7}
$$

Where $\{F\}_e$ represents external forces on the element and Δ is the area of the element. If the corresponding equations for all the elements are assembled in a global relation, the following expression gives the relation of forces versus displacements:

$$
\{F\} = [K]\{u\} \tag{3.8}
$$

In the last equation [K] is the global stiffness matrix. If external forces are known then the linear equations for displacements can be solved. Strains and stresses are calculated by Eqs. 3.4 and 3.5, respectively.

The example above pertains to small strain elasticity; plasticity and especially large deformations require more effort. Non-linearity may also be introduced. If a coupled analysis is to be followed the equations presented must include more variables. Generally speaking, machining requires non-linear and dynamic models so that it can be adequately simulated. All these are treated in the following paragraphs, focused on theories already used in machining.

If the models are non-linear and dynamic, how is time integration treated? There are two different time integration strategies used in the case of these problems, namely implicit and explicit schemes. The explicit approach determines the solution of the set of finite element equations by using a central difference rule to integrate the equations of motion through time. The equations are reformulated and they can be solved directly to determine the solution at the end of the increment, without iteration. The method is dynamic; it uses a mass matrix and computes the change in displacements from acceleration. On the other hand, the implicit method is realized by solving the set of finite element equations, performing iterations until a convergence criterion is satisfied for each increment. The length of the time step is imposed by accuracy requirements. In the implicit method the state of a finite element model at time $(t + \Delta t)$ is determined based on data at time $(t + \Delta t)$, while the explicit method solves the equations for $(t + \Delta t)$ based on data at time t. Both implicit and explicit methods have been used in cutting simulation [6, 7]. There are some papers that elaborate on the use of implicit or explicit techniques that give more information on the matter, with examples including manufacturing processes [8–10].

All the above may be confusing. *How can a modeler be aided?* It is true that modeling with FEM is not at all trivial. However, a solution would be to use a commercial FEM program. These programs usually have a CAD system implemented or can import files from the most widely used CAD programs, for geometry design, provide many element types, have mesh generation programs, include tools to apply boundary conditions, define contact conditions, perform coupled analysis, have non-linear capabilities and are supplied with automatic equation solvers. Most of them have a companion program for treating the results of the analysis. Furthermore, renown FEM software are expected to provide user support, including documentation, technical support, training courses and updates; a large user community is an additional benefit. On the other hand, programs such as these may be quite large and with numerous options. The user may be once again confused and produce models of limited value or misinterpret the results of the analysis, especially if the user lacks physical understanding needed to prepare a model. It would be wise to always validate model results, either with experiments or other models.

Furthermore, the modeler should be able to make a selection of a program that is suitable for the categories of problems to be faced. Not all programs have universal application, e.g. some perform only mechanical analysis and so coupled analysis is not carried out. Note also that usually the codes provided, which are undisclosed to users and cannot be altered in respect to this feature, are either implicit or explicit; the choice is not up to the user. Finally, commercial FEM are always accompanied with a cost that needs also to be taken into account. In Sect. 3.3.3 some commercial programs used in metal machining are presented.

3.2 Finite Element Modeling of Machining Considerations

In this section some aspects of FEM that are essential in order to provide realistic models and simulations for metal cutting processes are presented. This involves, among other topics, the presentation of the model formulation, the application of modeling strategies on mesh generation, the determination of boundary conditions and the modeling of workpiece material and tool-chip interface. Although this discussion involves techniques generally used in FEM, the application in machining is only documented here. The selection of the most appropriate of these features often determines the quality of the analysis that is carried out. Depending on selecting the "correct" parameters some important features are influenced and determined, such as cutting forces, temperatures and chip morphology. For example in FEM cutting models the workpiece material and the friction model in the tool-chip interface are considered of great importance for the outcome of the analysis. Many simulations are concerned only with the determination of these two factors in order to provide as accurate as possible predictions; the results may be further used for the determination of tool wear and surface quality characteristics.

Actually, there are no "correct" parameters in the sense that there is an on going research regarding the parameters that will prove to provide better results; by this it is meant results that can provide models to be used with several workpiece materials, cutting tools, processing conditions and provide simulation results with the minimum discrepancy from experiments carried out with the same set-ups. In each paragraph, references are given to show how the topics reported here are implemented by various researchers, how they justify the use of one or another feature and report their performance. It is up to the modeler to decide which features to implement in his model and in which way.

3.2.1 Model Formulation

In this paragraph the numerical formulations used in metal cutting FEM models are discussed. So far three types of analysis have been proposed, namely Eulerian, Lagrangian and the newer Arbitrary Lagrangian-Eulerian (ALE) analysis.

In the Eulerian approach the finite element mesh is spatially fixed and covers a control volume. The material flows through it in order to simulate the chip formation. This implies that the shape of the chip, shear angle and the contact conditions must be a priori known, derived from experiments, or assumed. An iterative procedure is used for the convergence of variables and chip geometry is updated. The element sides that are the boundaries of the chip that are adjacent to the rake face and far from the rake face of the tool are repositioned to be tangential to the cutting position. However, strains are derived from the integration of strain rates along stream lines; this cannot be used for the simulation of discontinuous chips.

In the Lagrangian approach the elements are attached to the material. The material is deformed due to the action of the cutting tool and so is the mesh. This way there is formation of the chip due to deformation from the tool. Unconstrained material flow in Lagrangian formulation allows for simulations from incipient chip formation to steady-state conditions and modeling of segmented chips besides the continuous one. In an explicit approach the displacement of the workpiece and the attached mesh, is a function of the time step and can be related to the material removal rate; in an implicit formulation the size of the time step has no influence on the stability of the solution. Furthermore, several models that depend on strain, strain rate and temperature have been applied for the workpiece material.

A disadvantage of the Lagrange formulation is connected to the large mesh deformation observed during the simulation. Due to the attachment of the mesh on the workpiece material, the mesh is distorted because of the plastic deformation in the cutting zone. Such severe distortions of the mesh may result in the failure of the model as they cannot be handled by the elements applied in the mesh. Pre-distorted meshes [11] and re-meshing techniques are applied in order to overcome these problems [12]. Furthermore, for the formation of the chip, a chip separation criterion in front of the tool edge is applied. This procedure can be quite thorny; it has been the topic of several papers and no generally accepted criterion is adopted. The latest development in the Lagrangian formulation, an updated Lagrangian analysis, has overcome the disadvantage of a chip separation criterion by applying continuous re-meshing and adaptive meshing, dealing at the same time with the mesh distortion; the above are thoroughly discussed in forthcoming paragraphs.

Summarizing a comparison between Eulerian and Lagrangian techniques it can be stated that the Eulerian formulation needs no re-meshing since there is no element distortion involved in the analysis and requires no chip separation criterion because the course of the chip is predetermined. The computational time in such models is reduced due to the few elements required for modeling the workpiece and the chip and simple procedures are used in the relative software. This analysis is suitable for the simulation of steady-state cutting, when the incipient stages of chip formation are not of interest and with continuous chip since no chip breakage criterion can be incorporated in the model for the simulation of discontinuous chip formation. This technique was used in the past, mainly in the first FEM models that appeared for metal cutting, e.g. in [13]. Although it is still used today [14], it is considered that it does not correspond to the real deformation procedure encountered in real metal cutting processes, as the chip thickness, a major outcome of the process cannot be assumed physically [15]. On the other hand, the Lagrangian and the updated Lagrangian formulation can produce non steady-state models with chip breakage considered; a lot of updated Lagrangian models have enriched the relative literature, such as [16–19]. Although the updated Lagrangian formulation adds considerably to the required calculation time, the advances in computers have made it possible to reduce the time needed for such an analysis to acceptable levels. However, two new aspects of machining modeling are introduced that their application is controversial and needs to be further studied, namely the use of a chip separation criterion and adaptive meshing.

The arbitrary Lagrangian-Eulerian formulation has also been proposed with the aim to combine the advantages of the two aforementioned methods [20–22]. This method uses the operator split procedure. The mesh is neither fixed nor attached to the material. Instead, it is allowed to arbitrarily move relative to the material with the total displacement being the sum of a Lagrangian displacement increment and an Eulerian displacement increment. A Lagrangian step is used in the procedure for the material flow at the free boundaries so that chip formation is the result of material deformation, thus mesh displacement in this step is associated with deformation. Then, in an Eulerian step, the reference system is suitably repositioned to compensate for the distortions during deformation, thus mesh displacement in this step is connected with numerical benefits. The procedure involves small time increments and it does not alter elements and connectivity of the mesh. Additionally, no separation criterion or extensive re-meshing is required. As a result, an ALE mesh is expected to be less distorted and more regular in comparison to a Lagrangian mesh. The drawbacks of the ALE formulation are the re-mapping of state variables, which may be performed inaccurately, and the need for a complete re-meshing [15].

3.2.2 Mesh, Elements, Boundary Conditions, Contact

The initial mesh of the workpiece is very significant for the results the model will provide. The convergence of the numerical procedure and the accuracy of the predicted variables depend on it. The obvious is that the mesh must be able to represent accurately the workpiece geometry and be able to handle the analysis to be performed. Structured and unstructured mesh generation procedures have been developed for the arrangement of the elements in the mesh and their individual geometry but there is not only one way to devise a representation of a continuum with finite elements. The size, number and type of the elements used in the mesh play a significant role on the simulation outcome as well.

As a rule, a large number of small sized elements increases accuracy but also increases computational time. There is a threshold beyond which further increase in the number of the elements will significantly increase the time of the analysis with marginal gain in accuracy. Usually, coarser meshes are used for testing a model and a finer mesh is applied when the model is checked. Another technique is to identify the regions that are of more interest for the analysis. Finer meshes may be used in these regions that are combined with larger ones in the other regions. In machining the action takes place in the primary and secondary deformation zones; the mesh in these parts of the workpiece is expected to be denser in order to obtain better geometry of the chip and also be able to cope with the strains, strain rates and temperature gradients expected there. These parameters are incorporated in the analysis by a material model suitable for thermal, elastic and plastic effects to be accounted for; material modeling is the topic of the next paragraph. In Fig. 3.2 an example of finer discretization in the regions where primary and secondary deformation zones are anticipated are shown. Note also that the chip has finer mesh than the workpiece, except the deformation zones and tool tip, so that the mesh follows chip shape more accurately.

An element with a compact and regular shape is expected to perform better, i.e. as the aspect ratio of an element increases it loses accuracy. Low order elements, often with formulations to avoid volumetric locking behavior that can halt the analysis due to large incompressible plastic strains in the cutting area, are widely

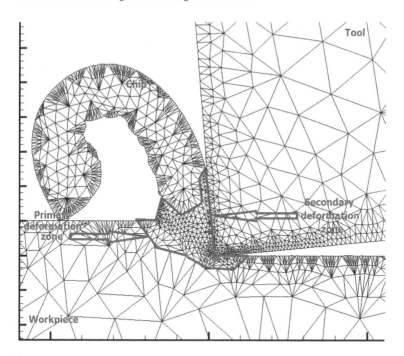

Fig. 3.2 Example of discretization

used due to their simplicity. The influence of numerical parameters on the modeling of orthogonal cutting is the topic of [23]. In the relevant literature quadrilateral elements with 8 [24–26] and 9 nodes [27], enhanced 4-noded elements [28–31] and triangular elements [32, 33] with 6 nodes [12] can be found.

The boundary conditions applied in the initial mesh may differ. In the case of workpiece and tool different approaches have been proposed. Some researchers apply boundary conditions that allow the tool to advance towards the non-moving workpiece while others do just the opposite, considering the tool to be constrained from moving in the x-axis. In any case the result is equivalent; the tool and the workpiece have a relative movement to each other equal to the cutting speed. Contact and contact detection between chip and tool is of great importance in machining modeling. The most common algorithms used for solving contact problems are the penalty approach and the Lagrangian multipliers approach. Other procedures such as the augmented Lagrangian technique and the perturbed Lagrangian method are reported [34].

Of interest is the way thermo-mechanical coupling is considered. In cutting processes heat generation originates from the two deformation zones, i.e. the primary and the secondary, due to inelastic and frictional work. The heat is conducted into the tool and chip and transferred away from the chip to the environment or the cutting fluid by convection. The above are either modeled by heat sources at the heat generation regions or usually with material and tribological models that are functions of mechanical and thermal behavior with strain, strain rate and temperature. The associated strain hardening and thermal softening is interpreted to non-linear analysis.

In [12] a staggered procedure is adopted for coupling thermal and mechanical equations. There are two different meshes, one for the mechanical and one for the thermal model, which exchange information. A mechanical step is taken first with constant temperature and heat generation is computed. Then it is transferred to the thermal mesh. Temperatures are re-computed and transferred to the mechanical mesh to be inserted to the thermal softening model used in the analysis. In another approach, all heat generated by deformation and friction is kept inside the mesh, external boundaries of workpiece, chip and tool are insulated, causing temperature rise [23, 35–37]. This adiabatic approach can be employed for low diffusivity materials in high speed machining [34]. If external boundaries are not adiabatic then heat convection takes place through natural convection or forced convection by the cutting fluid, for wet machining. Two difficulties arise in this case: the long standing problem of whether the cutting fluid actually makes its way between chip and tool and what are the actual thermal characteristics of the cutting fluid, e.g. heat transfer coefficient, in this region. Heat loss due to radiation is ignored, as it is considered negligible. Thermal contact, i.e. the numerical technique to model heat transfer between chip and tool, is realized by several approaches such as the heat conduction continuity, the constant factor, two semi-infinite bodies and the thermal equilibrium approaches [38]. It is worth noticing that for commercial FEM software various options and tools on the above are provided.

3.2.3 Material Modeling

Material modeling in machining is of great importance. Especially the flow properties of the workpiece material and the corresponding equations that are included into FEM have been extensively studied. These constitutive equations describe the flow stress or instantaneous yield strength at which work material starts to plastically deform or flow; the elastic strains are much lower than plastic strains in metal cutting and so workpiece material flows plastically into the cutting zone. The constitutive models presented in the literature are mainly elastic-plastic [30, 39–43], elastic-viscoplastic [29, 32], rigid-plastic [44–46] and rigid-visco-plastic [47–50].

Machining conditions subject workpiece material to high levels of strain, strain rate and heat which greatly influence flow stress. In the primary zone strain and temperature ranges from 1 to 2 and 150–250 °C respectively and in the secondary deformation zone from 3 to much higher and 800–1200 °C, while strain rates reach values of up to $2 \times 10^4 \, \text{s}^{-1}$ and $10^5 \, \text{s}^{-1}$ in the two zones [51]. If σ is stress, ε is plastic strain, $\dot{\varepsilon}$ is plastic strain rate and T is temperature, a complete constitutive equation is in the form:

$$f(\sigma) = \sigma(\varepsilon, \dot{\varepsilon}, T) \tag{3.9}$$

The problem is the lack of data for high stresses, strain rates and temperatures as the ones encountered in machining. In many cases the constitutive data are taken from standard tension tests that are not sufficient for machining processes. Dynamic experimental material tests such as Split Hopkinson Pressure Bar (SHPB) impact testing is employed. Samples are deformed under high speed compression with strain rates of up 10^5 s^{-1} and temperatures of up to 700 °C. However, the results are not sufficient for the deformation behavior of metals, especially in high speed machining; values beyond test results are calculated by interpolation. Astakhov and Outeiro criticized the use of SHPB results in machining [15]. They argue that the available data are not from specialized laboratories, generally speaking SHPB requires special equipment; high strain rates in metal cutting is a myth [52], metal cutting is a cold working process, although the chip only is of high temperature; finally, it not clear how to correlate uniaxial impact testing results of SHPB with materials that are triaxially stressed, as in metal cutting. Other tests used are torsion tests, compression ring tests and projectile impact tests [53].

Although many constitutive equations have been employed for the case of metal cutting, some are discussed here. The first is the relation by Usui, Maekawa and Shirakashi [54, 55]:

$$\sigma = B\left[\frac{\dot{\varepsilon}}{1000}\right]^M e^{-kT}\left[\frac{\dot{\varepsilon}}{1000}\right]^m \left\{\int_{Path} e^{kT/N}\left[\frac{\dot{\varepsilon}}{1000}\right]^{-m/N} d\varepsilon\right\}^N \tag{3.10}$$

In this equation B is the strength factor, M is the strain rate sensitivity and n the strain hardening index, all functions of temperature T, and k and m are constants. The integral term accounts for the history effects of strain and temperature in relation to strain rate. In the absence of these effects, Eq. 3.10 is reduced to [56]:

$$\sigma = B\left[\frac{\dot{\varepsilon}}{1000}\right]^M \varepsilon^N \tag{3.11}$$

Oxley suggested a relation for carbon steel as [57]:

$$\sigma = \sigma_1 \varepsilon^n \tag{3.12}$$

with σ_1 the material flow stress for $\varepsilon = 1$ and n is the strain hardening exponent. Both are functions of temperature, which is velocity modified as:

$$T_{\text{mod}} = T[1 - 0.09 \log(\dot{\varepsilon})] \tag{3.13}$$

for the combined effect of temperature and strain rate.

Among the most used material models is the Johnson-Cook model [58]. The equation consists of three terms the first one being the elastic-plastic term to represent strain hardening, the second is viscosity, which demonstrates that

Table 3.1 Material models in metal cutting modeling

Model	Constitutive equation	Reference
Usui et al.	$\sigma = B\left[\frac{\dot{\varepsilon}}{1000}\right]^M e^{-kT}\left[\frac{\dot{\varepsilon}}{1000}\right]^m\left\{\int_{Path} e^{kT/N}\left[\frac{\dot{\varepsilon}}{1000}\right]^{-m/N} d\varepsilon\right\}^N$	[54, 55]
Oxley	$\sigma = \sigma_1 \varepsilon^n$	[57]
Johnson-Cook	$\sigma = (A + B\varepsilon^n)\left[1 + C\ln\left(\frac{\dot{\varepsilon}}{\dot{\varepsilon}_o}\right)\right]\left[1 - \left(\frac{T-T_a}{T_m-T_a}\right)^m\right]$	[58]
Zerilli-Armstrong	$\sigma = C_o + C_1 \exp[-C_3 T + C_4 T \ln(\dot{\varepsilon})] + C_5\varepsilon^n$ $\sigma = C_o + C_2\varepsilon^n \exp[-C_3 T + C_4 T \ln(\dot{\varepsilon})]$	[65]

material flow stress increases for high strain rates and the temperature softening term; it is a thermo-elastic-visco-plastic material constitutive model, described as:

$$\sigma = (A + B\varepsilon^n)\left[1 + C\ln\left(\frac{\dot{\varepsilon}}{\dot{\varepsilon}_o}\right)\right]\left[1 - \left(\frac{T - T_a}{T_m - T_a}\right)^m\right] \qquad (3.14)$$

where $\dot{\varepsilon}_o$ is the reference plastic strain rate, T_α the ambient temperature, T_m the melting temperature and A, B, C, n and m are constants that depend on the material and are determined by material tests [59, 60] or predicted [61]. The influence of the Johnson-Cook constants on the outcome of machining modeling was investigated [62] and was found that FEM results are sensitive to these inputs, which in turn are strongly related to the test method used to derive the constants. On the other hand the results from a test method can be fitted to different constitutive equations and the selection of the material model can influence the predicted results [63, 64].

Zerilli and Armstrong developed a constitutive model based on dislocation-mechanics theory and considering crystal structure of materials [65]. They suggested two different models, one for body cubic centered (BCC) and one for face cubic centered (FCC) lattice structure, respectively:

$$\sigma = C_o + C_1 \exp[-C_3 T + C_4 T \ln(\dot{\varepsilon})] + C_5\varepsilon^n \qquad (3.15)$$

$$\sigma = C_o + C_2\varepsilon^n \exp[-C_3 T + C_4 T \ln(\dot{\varepsilon})] \qquad (3.16)$$

where C_i, i = 0–5, and n are material constants determined experimentally, e.g. by the SHPB method [66]. Table 3.1 summarizes the constitutive models that are commonly found in metal machining modeling.

In most analyses performed, cutting tool is considered as a rigid body, although exceptions exist [13, 29, 46, 67]. The tool is not deformed; however, thermal analysis for the determination of the temperatures, especially in the tool tip, can be carried out. If coatings are also modeled, they are modeled as elastic materials and only heat transfer and elastic material properties are needed [68].

3.2.4 Friction Modeling

Friction modeling in the secondary deformation zone, at the interface of the chip and the rake face of the tool is of equal importance to the workpiece material modeling presented in the previous paragraph. It is important in order to determine cutting force but also tool wear and surface quality. Once again the detailed and accurate modeling is rather complicated. Many finite element models of machining assume that it is a case of classical friction situation following Coulomb's law; frictional sliding force is proportional to the applied normal load. The ratio of these two is the coefficient of friction μ which is constant in all the contact length between chip and tool. The relation between frictional stresses τ and normal stresses may be expressed as:

$$\tau = \mu\sigma \tag{3.17}$$

However, as the normal stresses increase and surpass a critical value, this equation fails to give accurate predictions. From experimental analysis it has been verified that two contact regions may be distinguished in dry machining, namely the sticking and the sliding region. Zorev's stick-slip temperature independent friction model is the one commonly used [69]. In this model there is a transitional zone with distance ℓ_c from the tool tip that signifies the transition from sticking to sliding region. Near the tool cutting edge and up to ℓ_c, i.e. the sticking region, the shear stress is equal to the shear strength of the workpiece material, k, while in the sliding region the frictional stress increases according to Coulomb's law.

$$\tau = \begin{cases} k, 0 \leq \ell \leq \ell_c \\ \mu\sigma, \ell > \ell_c \end{cases} \tag{3.18}$$

In machining other approaches, based on Zorev's model, have been reported that include the defining of an average friction coefficient on the rake face or different coefficients for the sliding and the sticking region. In another approach, the constant shear model assumes that the frictional stress on the rake face of the tool is equal to a fixed percentage of the shear flow stress or the workpiece material.

Usui, based on Zorev's model and experimental results [54] proposed a non-linear stress expression:

$$\tau = k\left[1 - \exp\left(-\frac{\mu\sigma}{k}\right)\right] \tag{3.19}$$

This equation approaches the sticking region part of Eq. 3.18 for large σ and the sliding part for smaller values. However, the mean friction stress on the tool rake face may differ from the frictional stress in the sticking region. Childs [70] proposed another model:

$$\tau = mk\left[1 - \exp\left(-\frac{\mu\sigma}{mk}\right)^n\right]^{1/n} \tag{3.20}$$

In the last equation, m and n are correction factors; the former ensures that at high normal stresses the frictional stresses do not exceed k and the latter controls the transition from sticking to sliding region. These coefficients can be obtained by split-tool tests.

Iwata et al. [44] proposed a formula where Vickers hardness is also included. This equation is a close approximation to Usui's model if (H_V/0.07) is replaced by ($m\tau$):

$$\tau = \frac{H_V}{0.07} \tanh \left(\frac{\mu\sigma}{H_V/0.07} \right) \tag{3.21}$$

Other models proposed are the ones from Sekhon to Chenot [71] and Yang and Liu [28]. The first one employs Norton's friction law and includes the relative sliding velocity between chip and cutting tool v_f. In this equation α is the friction coefficient, K is a material constant and p a constant that depends on the nature of the chip-tool contact. The second one relates frictional and normal stresses through a polynomial series. The fourth order polynomial approximates Eqs. 3.19 and 3.21. The aforementioned equations are:

$$\tau = -\alpha K \left\| v_f \right\|^{p-1} v_f \tag{3.22}$$

And

$$\tau = \sum_{k=0}^{4} \mu_k \sigma^k \tag{3.23}$$

The evaluation of friction models has been the topic of some publications. An updated Lagrangian model to simulate orthogonal cutting of low carbon steel with continuous chip was prepared [18]. In a reverse engineering approach, five different friction models were tested and the results were compared against experimental results to decide which friction model is the most suitable. The results were best when friction models with variable shear stress and coefficient of friction were incorporated with the finite element models. Furthermore, an ALE model was used to measure the influence of friction models on several parameters [72]. It was concluded that friction modeling affects thrust forces more than cutting forces. Furthermore, on the implementation of the stick-slip model it is concluded that a major disadvantage is the uncertainty of the limiting shear stress value. In another work [73], five different friction models were analyzed and the investigators concluded that mechanical result, e.g. forces, contact length, are practically insensitive to friction models, as long as the "correct" friction coefficient is applied, while on the other hand, friction modeling greatly affects thermal results. In [74] an improved friction law formulation is suggested where the constant friction coefficient is replaced by one which increases with plastic strain rate:

$$\mu = \mu_o(1 + \alpha\dot{\varepsilon}^p) \tag{3.24}$$

Another parameter, which is closely connected to friction and FEM modeling, the contact length, is analyzed in [75]. Several contact length models utilized in the

Table 3.2 Friction models in metal cutting modeling

Model	Equation	Reference
Coulomb	$\tau = \mu\sigma$	–
Zorev	$\tau = \begin{cases} k, 0 \leq \ell \leq \ell_c \\ \mu\sigma, \ell > \ell_c \end{cases}$	[69]
Usui	$\tau = k\left[1 - \exp\left(-\frac{\mu\sigma}{k}\right)\right]$	[54]
Childs	$\tau = mk\left[1 - \exp\left(-\frac{\mu\sigma}{mk}\right)^n\right]^{1/n}$	[70]
Iwata et al.	$\tau = \frac{H_V}{0.07}\tanh\left(\dfrac{\mu\sigma}{H_V/0.07}\right)$	[44]
Sekhon and Chenot	$\tau = -\alpha K\|v_f\|^{p-1}v_f$	[71]
Yang and Liu	$\tau = \sum\limits_{k=0}^{4} \mu_k\sigma^k$	[28]

prediction of contact length in machining are analyzed. It should be noted that several papers presume frictionless contact in the chip-tool interface. Finally, it is observed [15] that in several experimental data provided in the relevant literature, friction coefficients are well above the value of 0.577; above this value no relative motion at the tool-chip interface can occur [52]. It is assumed [76] that friction coefficients above 1 need the strongest levels of adhesion between asperities and the tool; these conditions may be encountered at the newly formed chip and at high temperatures as those in the chip-tool interface.

The friction models discussed in this paragraph are summarized in Table 3.2.

3.2.5 Chip Separation–Chip Breakage

As pointed out in Sect. 3.2.1, Lagrangian formulation based models, simulate chip generation either by plastic deformation considerations or by employing a chip separation criterion. The ideal would be to incorporate to the model the real physical mechanism of chip formation for a material being machined. It is generally thought that chip formation in ductile metal cutting involves only plastic deformation without any fracture. Many researchers prefer this approach as being more realistic, backed-up by the fact that microscopic observations of the cross-sectional areas of the chip revealed no evidence of a crack. It is argued that chip formation cannot be accomplished just by plastic deformation [77]. The fact that no crack is observed in the laboratory tests is attributed to crack stability rather than crack formation; large compressive stresses in the tool edge quench cracks or cracks have the same speed with the tool and cannot be seen [78]. Thus, the implementation of a separation criterion to simulate separation and fracture of the material is not only a modeling technique to overcome the problem of chip formation.

Most of the available models of machining pertain to continuous chip. To model shear localized chip a suitable damage criterion is needed to simulate chip breakage. These criteria are similar or the same to the ones used to describe the onset of chip formation. This is why chip separation and chip breakage criteria are discussed together in this paragraph.

The two main techniques for chip separation are node-splitting and element deletion techniques [79]. In the node-splitting case a chip separation plane is predefined and a separation criterion is applied. There are two types of criteria, namely geometrical and physical. A simple geometrical criterion is to prescribe a critical distance d_c between the tool tip and the nearest node on the cutting direction [80]. This criterion is easy to control and can be used for cutting tools without sharp edge but it cannot account for breakage outside the cutting line. Furthermore, different critical distances result in different plastic strain distribution [81]. Physical criteria use the critical value of a physical quantity to estimate the onset of separation, e.g. in the plastic strain criterion chip separated when the calculated plastic strain at the nearest node to the cutting edge reaches the critical value [82–84]. The disadvantage of this criterion is that node separation may propagate faster than cutting speed, "unzipping" the mesh in front of the cutting tool. Another physical criterion is connected to stress [85]. Based on the Johnson-Cook yield stress equation a critical strain to fracture criterion is used [23, 43].

For chip breakage, a fracture criterion based on the critical stress for brittle mode fracture and another based on effective plastic strain for ductile failure was introduced by Marusich and Ortiz [12]. Obikawa et al. [86] used a criterion on equivalent plastic strain for producing discontinuous chip; when the equivalent plastic strain exceeded fracture strain, crack nucleation and growth occurred.

An interesting review of the chip separation criteria can be found in [6]. The author's state that the criteria reviewed cannot simulate incipient cutting correctly. The matter remains controversial and more research needs to be carried out to improve separating and breakage criteria for metal machining. In Table 3.3 chip separation and breakage criteria are gathered.

3.2.6 Adaptive Meshing

In the Lagrangian formulation, the initial mesh is altered significantly due to plastic deformation and chip separation. The distorted mesh causes numerical errors and the solution is rapidly degraded; the Jacobian determinant becomes negative for severe distortion and the analysis is halted. A strategy to address this problem is to use pre-distorted meshes [11, 16, 43, 87, 93–95]. The advent of computers has made it possible to apply adaptive meshing techniques. During the simulation certain steps are taken, e.g. the size of the elements, the location of the nodes or the number of the elements changes so that a new mesh, applicable for the analysis is created; this procedure takes place periodically. Figure 3.3 shows an example of an adaptive meshing technique, through the steps of the analysis.

Adaptive meshing can take place as a re-meshing technique, where the existing distorted mesh is substituted by a new one. The refinement technique increases

Table 3.3 Chip separation and breakage criteria [34]

Criterion	Definition	References
Nodal distance	$d = d_{cr}$	[11, 16, 25, 31, 40, 41, 80, 86–89][a]
Equivalent plastic strain	$I_{cr} = \varepsilon$	[28, 29, 42, 82, 83, 90][a] [35, 37][b]
Energy density	$I_{cr} = \int \sigma : d\varepsilon$	[81, 91][a] [92, 93][b]
Tensile plastic work	$I_{cr} = \int (\sigma_1/\sigma_Y)d\varepsilon$	[47][a] [44][b] [45, 94][c]
Brozzo et al.	$I_{cr} = \int \left(2\sigma_1/3(\sigma_{1-\sigma_H})\right)d\varepsilon$	[48][c]
Oskada et al.	$I_{cr} = \int (\varepsilon + b_1\sigma_H + b_2)d\varepsilon$	[44][b]
Stress index	$f = \sqrt{(\sigma/\sigma_f)^2 + (\tau/\tau_f)^2}$	[25, 95, 96][a]
Maximum principal stress	$\sigma_f = \sigma_1$	[97][c]
Toughness	$\sigma_f = K_{1C}/\sqrt{2\pi\ell}$	[12][d]
Rice and Tracey	$\varepsilon_f = 2.48\exp(-3\sigma_H/2\sigma_Y)$	[12][c]
Obikawa et al.	$\varepsilon_f = \varepsilon_o - \alpha p/\sigma - \beta\dot{\varepsilon}/v_c$	[86][c]
Obikawa and Usui	$\varepsilon_f = -[0.075\ln(\dot{\varepsilon}/100)] - \sigma_H/37.8 + 0.09\exp(T/293)$	[98][c]
Johnson-Cook	$\varepsilon_f = [D_1 + D_2\exp(D_3\sigma_H/\sigma_Y)]$ $\times [1 + D_4\ln(\dot{\varepsilon}/\dot{\varepsilon}_o)][1 + D_5(T-T_a/T_m-T_a)]$	[42, 99][c] [23, 43][a]
Damage considerations	$\varepsilon_f = A\{\sigma_Y^2/2Er[2/3(1+v) + 3(1-2v)(\sigma_H/\sigma_Y^2)]\}^{-s} + \partial\varepsilon/\partial T(T - T_o)$	[7, 36][e]

[a] Chip separation along a pre-defined parting line/plane
[b] Chip breakage
[c] Steady-state analysis without actual chip separation or breakage
[d] Fracturing material and brittle-type fracture
[e] Multi-fracturing materials and chip breakage

mesh density by reducing mesh size and smoothing relocates nodes to provide more regular element shapes. Adaptive meshing improves the accuracy of the simulation but at a computational cost. This is attributed to the newer mesh being denser and thus more elements are involved in the analysis, but also to two other very important aspects of re-meshing, namely error and distortion metrics, for the assessment of the quality of the solution and transfer operators, that are responsible for transferring the variables of the old mesh to the new [101].

3.3 Finite Element Method in Machining Bibliography

In this paragraph a bibliographical review on publications related to FEM modeling is provided. A search in references reveals hundreds of papers published since the early 1970s on this topic. The extensive number of publications

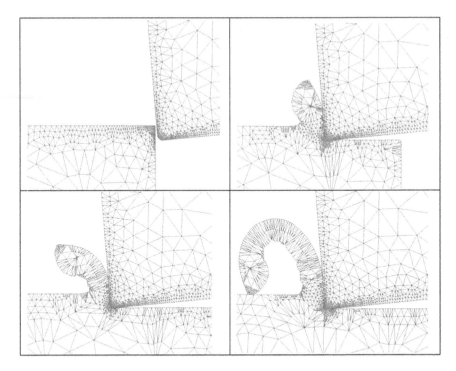

Fig. 3.3 Example of adaptive meshing [100]

pertaining to the application of FEM in metal cutting demonstrates the importance of the method in this area of application but at the same time it makes the tracking of innovation quite difficult. Such an amount of papers makes it difficult to cite all of them. In the next few lines some papers will be mentioned to outline the historical development of the application of the method in machining. In the next chapter more papers will be discussed as they are of importance to the topic of the paragraph they are included. Finally, next chapter is dedicated to the extensive discussion of case studies of machining models.

3.3.1 The First Three Decades: 1971–2002

The earliest chip formation study is included in the book of Zienkiewicz [102]. The presented model is a simple small strain elastic-plastic analysis with no friction between the chip and the tool. The chip is preformed and the tool is advanced towards it to simulate steady-state cutting conditions. Klamecki presented a three dimensional model that was limited to the first stages of chip formation [103]. Worth mentioning is an early pioneering work of Shirakashi and Usui on orthogonal cutting simulation [104]. They developed a computational

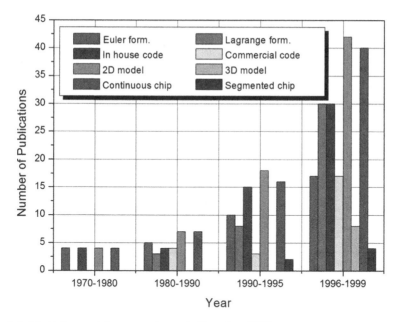

Fig. 3.4 Machining modeling research trends (Data from [79])

method called Iterative Convergence Method (ICM). For the creation of the chip they used propagation of a small crack in front of the tool. The first models that appeared on the simulation of metal machining, and the majority of the work performed so far, pertain to two dimensional orthogonal cutting plain-strain models. In the early stages of FEM modeling of machining Eulerian formulation was preferred and some researchers still use it [13, 70, 105, 106]. However, the Lagrangian formulation is more often encountered in metal machining simulations [16, 81, 87, 89, 107], and so is the updated Lagrangian formulation [12, 17, 45, 46, 108]. The first ALE formulation models also begin to appear [20, 21]. Finally, 3D FEM modeling is used by Cerreti et al. [109] in order to simulate orthogonal and oblique cutting conditions in turning of aluminum and steel. These citations show the trends in FEM metal machining modeling. In Fig. 3.4 the same trends are shown, with data taken from [79]. In the first decade 2D continuous chip Eulerian models developed with in-house FEM codes are reported. In the next 10 years, Lagrange models and commercial FEM software appear. Between 1990 and 1995, the number of publications, in reference to machining, increase rapidly while segmented chip models are cited. In the last time period, the number of relevant to machining publications is larger than all the previous years together. Although 3D models are constructed, still 2D models with continuous chip are the majority of cited papers.

In 2001 CIRP conducted a survey among its members on the available predictive performance models of machining operations [110]. The survey indicated that 28 research groups had developed some kind of predictive model for a

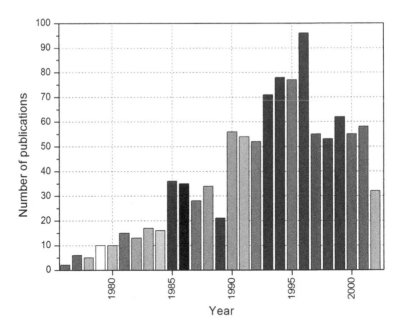

Fig. 3.5 Publications/year on machining FEM models from 1976 to 2002 (Data from [111, 112])

variety of 16 different operations. More specifically, 75 % of the groups were involved with models for turning with plane face lathe tools, 53.5 % with face milling and 50 % with drilling with twist drills; the most sought after parameters being the force components with almost 70 % of the researches possessing a model able to predict it. However, for turning, 39.3 % of the groups responded that they used an empirical model, 52.2 % an analytical model, 26.5 % a mechanistic model and only 11 % a FEM or Artificial Intelligence model. The number of publications relevant to cutting is also low. For instance, for 1991–1994 the average number was 22.5 papers/year while for 1995–2000 the number dropped to 12.5 papers/year; however, this was attributed to the high rejection rates, in excess of 50 %, in the second period.

Mackerle prepared two papers where publications from 1976 to 1996 and 1996 to 2002 are collected [111, 112]. In these bibliographies 1,047 papers from journals, conferences and theses in total on finite element modeling and simulation of machining are presented. In Fig. 3.5, the number of publications for each year is presented. It can be seen that up to 1996 the publications exhibit an increasing tendency, especially after 1990. In the same papers by Mackerle, the publications are divided into 9 categories, as seen in Fig. 3.6.

Category 2 is further divided into 7 subcategories, denoted (a–f) in Fig. 3.6, one for each machining processes. However, subcategory (a) is underestimated since turning publications may belong in other categories, too.

3.3.2 The Last Decade: 2002–2012

In the last 10 years, machining models still continue to deal with updated Lagrangian [18, 19] and ALE formulation [72, 113–114]. Furthermore, the investigations on material and friction modeling and chip breakage continue with irreducible interest, as pointed out in the relevant paragraphs [18, 114]. All the proposed models in the relevant literature deal mainly with features such as chip morphology, cutting forces, temperatures, surface integrity, residual stresses and tool wear for the machining of steels and other metals such as aluminum and titanium. Most of the relative work examines turning but milling and drilling is considered as well, while models and simulations of 3D nature are not so common. The reasons behind this are that for 3D modeling and simulation, the degree of complexity and computational power required, are increased. 3D FEM modeling was used by Ceretti et al. in order to simulate orthogonal and oblique cutting conditions in turning of aluminum and steel [109]. Aurich and Bil [115] presented a 3D model that produced serrated chip, while Attanasio et al. [116] and Özel [117] dealt with tool wear. Summarizing, the trends of machining simulation indicate that the interest has shifted over to simulation of processes other than turning [118, 119], 3D models [120], machining of hard-to-machine materials, e.g. Titanium and Titanium alloys [121], high speed hard machining [122], precision machining [123] and micro-machining [124]. Some of these kinds of modeling are discussed in Chap. 4 with examples and a review of the relevant literature is provided there.

3.3.3 FEM Software

Models of the early publications were constructed by FEM codes made in-house by the researchers. For the past 20 years a wide range of commercial FEM packages became available. These programs have been widely accepted by researcher since they can simplify the overall procedure of model building. Commercial FEM add to the quality and accuracy of the produced models. These programs are made by specialists who have tested them and have implemented features and procedures to accelerate the slow procedure of model building. Most of the software have mesh generation programs, easy to use menus for applying boundary conditions, contact algorithms, automatic re-meshing, material databases etc. Some researchers, however, remain skeptical due to limitations a model can impose, e.g. a model may only be able to solve a problem implicitly or explicitly.

Regarding machining, FEM codes that have been used for the simulation of machining include: Abaqus [43, 79, 125], FORGE 2 [32], NIKE-2D [107], DEFORM 2D/3D [18, 49, 62, 126], I-FORM 2 [78], MSC.Marc [41], LS-DYNA [127] and AvantEdge [19, 78, 122, 123, 128, 129]. Some of these programs are

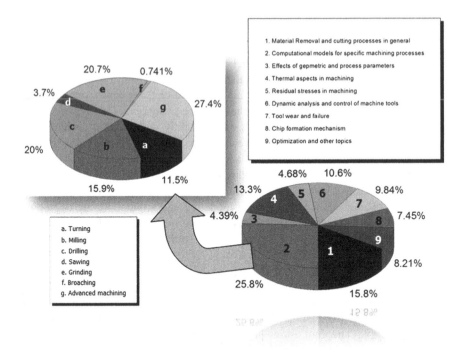

Fig. 3.6 Main topics of the FEM models (Data from [111, 112])

general purpose software, i.e. MSC.Marc and ABAQUS, or specific purpose programs for machining or other uses. LS-DYNA is a program used mainly in crashworthiness analysis and forming problems, DEFORM 2D is a program for forming processes, which has a machining module to accommodate turning, milling, boring and drilling operations and AdvantEdge is a FEM program that is only used to simulate machining. It can simulate orthogonal and oblique cutting, 3D modeling of turning, milling, drilling, boring and taping, and 2D modeling of micromachining among others. AdvantEdge is a Lagrangian, explicit, dynamic code which can perform coupled thermo-mechanical transient analysis. The program applies adaptive meshing and continuous re-meshing automatically. The drawback of this software is that some parameters are fixed and the user cannot intervene, e.g. friction coefficient is constant in the tool-chip interface.

Bil, Kılıç and Tekkaya have compared models from three different commercial software, namely, MSC.Marc, DEFORM 2D and AdvantEdge [130]. In all three cases an orthogonal plane strain model is constructed but some features of the models are quite different, e.g. MSC.Marc and DEFORM 2D are implicit codes while AdvantEdge is explicit, chip separation in MSC.Marc and AdvantEdge is through re-meshing and in DEFORM 2D a damage criterion and element deletion are applied. Other differences pertain to friction model, element types and material model. The results of the models are compared to experimental results and not a good agreement in all the parameters was found. Firstly, the material data are not obtained for high strain rates such as those encountered in machining but are extrapolated and need to be improved. Friction modeling needs to be tuned in order to provide more reliable results; although cutting forces are in good agreement with experimental ones, thrust forces are more accurately predicted with larger friction parameters. The plain Coulomb friction model, used in AdvantEdge, is not adequate for providing good approximations of the forces. However, it is argued that a more reliable chip separation criterion needs to be proposed. The authors state that, although chip formation by re-meshing provides better results, it is based on the misconception of crack generation in the material near the tool tip. The technique resembles crack generation because of the way the new mesh is formed after re-meshing.

It is argued and backed-up with experiments that the discrepancies between modeling and experimental results lay with the materials and the conditions and not with the failure of software to simulate machining. It is agreed that cutting and thrust forces are not correctly predicted at the same time, the latter being underestimated. This can be corrected by altering friction parameters or as suggested by Childs [131–133] by incorporating in the material model the effect of yield delay, a phenomenon taking place when machining carbon steels at elevated speeds.

3.4 Concluding Remarks

In Sects. 3.2.1–3.2.6 the parameters to be taken into account in order to construct a FEM model for machining are discussed and the available options for each case are laid down. It is up to the modeler to incorporate some or all of the parameters in his model and also try to figure out which option for which parameter will work better for the at hand problem. It needs to be decided what mesh will be applied on the workpiece geometry, both size and shape, what kinds of elements are to be used, what boundary conditions are to be applied and how surface contacts will be modeled. Then, the formulation must be decided, i.e. Eulerian, Lagrangian or ALE and if the problem will be solved implicitly or explicitly. A real puzzle must be solved in connection with the material and friction model to be used and whether adaptive meshing or a chip separation—and which—will be applied on the model. If the analysis is to considered 3D and special care is needed for simulating a chip other than a continuous one, complicates the process and adds considerably to computational time. Generally speaking, there is no material and friction model or chip separation criterion that is generally accepted. Furthermore, it is argued by many investigators that these parameters can significantly alter the model results. Finally, the software used for carrying out the analysis has some special characteristics of its own that affect the numerical solution. Commercial FEM software is, in many cases, important assistance.

Providing a reliable and physically sound model is not easy at all. One must have a strong background on the problem dealt with FEM, in this case machining, but also on FEM method as well. One simple rule could be to start simple with the model and anticipate the results; then revise the model so that it includes more detail both in a physical respect, e.g. add material properties parameters that are acquired by a proper method for the specific material, and in numerical respect, e.g. use more elements or more focused mesh. Sometimes, validation of the models is used in order to fine tune the model, which probably means that a model parameter is violating a physical law, e.g. too big friction coefficient. However, it is argued by most researchers that the underlying phenomena of chip formation are not fully understood yet. Validation is also a hard task; the equipment needed to make measurements of model input parameters, e.g. friction coefficient and material constants in Johnson-Cook model, and model outputs, e.g. cutting forces and temperatures, usually require sophisticated equipment. Furthermore, workshop conditions cannot be identical to the simulation ones.

However, FEM is considered the best option, in comparison to other methods, to provide reliable results, especially when combined with powerful computers. The accumulated experience on the method is also an advantage. The vast number of publications on FEM modeling of machining, described in 3.3, proves these statements.

References

1. Dixit PM, Dixit US (2008) Modeling of metal forming and machining processes by finite element and soft computing methods. Springer, UK
2. Klocke F, Beck T, Hoppe S, Krieg T, Müller N, Nöthe T, Raedt HW, Sweeney K (2002) Examples of fem application in manufacturing technology. J Mater Process Technol 120:450–457
3. Mamalis AG, Manolakos DE, Ioannidis MB, Markopoulos A, Vottea IN (2003) Simulation of advanced manufacturing of solids and porous materials. Int J Manuf Sci Prod 5(3):111–130
4. Cook RD, Malkus DS, Plesha ME (1989) Concepts and applications of finite element analysis, 3rd edn. Wiley, New York
5. Bathe K-J (1996) Finite element procedures. Prentice Hall, USA
6. Huang JM, Black JT (1996) An evaluation of chip separation criteria for the fem simulation of machining. J Manuf Sci Eng 118:545–554
7. Owen DRJ, Vaz M Jr (1999) Computational techniques applied to high-speed machining under adiabatic strain localization conditions. Comput Methods Appl Mech Eng 171:445–461
8. Lindgren LE, Edberg J (1990) Explicit versus implicit finite element formulation in simulation of rolling. J Mater Process Technol 24:85–94
9. Sun JS, Lee KH, Lee HP (2000) Comparison of implicit and explicit finite element methods for dynamic problems. J Mater Process Technol 105:110–118
10. Harewood FJ, McHugh PE (2007) Comparison of the implicit and explicit finite element methods using crystal plasticity. Comput Mater Sci 39:481–494
11. Shih AJ (1996) Finite element analysis of rake angle effects in orthogonal metal cutting. Int J Mech Sci 38:1–17
12. Marusich TD, Ortiz M (1995) Modelling and simulation of high-speed machining. Int J Numer Meth Eng 38:3675–3694

13. Strenkowski JS, Carrol JT III (1986) Finite element models of orthogonal cutting with application to single point diamond turning. Int J Mech Sci 30:899–920

14. Dirikolu MH, Childs THC, Maekawa K (2001) Finite element simulation of chip flow in metal machining. Int J Mech Sci 43:2699–2713

15. Astakhov VP, Outeiro JC (2008) Metal cutting mechanics, finite element modelling. In: Davim JP (ed) Machining: fundamentals and recent advances. Springer, UK

16. Shih AJ (1995) Finite element simulation of orthogonal metal cutting. ASME J Eng Ind 117:84–93

17. Bäker M, Rösler J, Siemers C (2002) A finite element model of high speed metal cutting with adiabatic shearing. Comput Struct 80:495–513

18. Özel T (2006) The influence of friction models on finite element simulations of machining. Int J Mach Tools Manuf 46:518–530

19. Maranhão C, Davim JP (2010) Finite element modelling of machining of aisi 316 steel: numerical simulation and experimental validation. Simul Model Pract Theory 18:139–156

20. Olovsson L, Nilsson L, Simonsson K (1999) An ALE formulation for the solution of two-dimensional metal cutting problems. Comput Struct 72:497–507

21. Movahhedy MR, Altintas Y, Gadala MS (2002) Numerical analysis of metal cutting with chamfered and blunt tools. Transactions of the ASME: J Manuf Sci Eng 124:178–188

22. Arrazola PJ, Özel T (2008) Numerical modelling of 3-D hard turning using arbitrary Eulerian Lagrangian finite element method. Int J Mach Mach Mater 3:238–249

23. Barge M, Hamdi H, Rech J, Bergheau J-M (2005) Numerical modelling of orthogonal cutting: influence of numerical parameters. J Mater Process Technol 164–165:1148–1153

24. Joshi VS, Dixit PM, Jain VK (1994) Viscoplastic analysis of metal cutting by finite element method. Int J Mach Tools Manuf 34:553–571

25. McClain B, Batzer SA, Maldonado GI (2002) A numeric investigation of the rake face stress distribution in orthogonal machining. J Mater Process Technol 123:114–119

26. Fihri Fassi H, Bousshine L, Chaaba A, Elharif A (2003) Numerical simulation of orthogonal cutting by incremental elastoplastic analysis and finite element method. J Mater Process Technol 141:181–188

27. Tyan T, Yang WH (1992) Analysis of orthogonal metal cutting processes. Int J Numer Methods Eng 34:365–389

28. Yang X, Liu CR (2002) A new stress-based model of friction behaviour in machining and its significant impact on the residual stresses computed by finite element method. Int J Mech Sci 44:703–723

29. Kishawy HA, Rogers RJ, Balihodzic N (2002) A numerical investigation of the chip-tool interface in orthogonal machining. Mach Sci Technol 6:397–414

30. Ohbuchi Y, Obikawa T (2005) Adiabatic shear in chip formation with negative rake angle. Int J Mech Sci 47:1377–1392

31. Mamalis AG, Branis AS, Manolakos DE (2002) Modelling of precision hard cutting using implicit finite element methods. J Mater Process Technol 123:464–475

32. Ng E-G, Aspinwall DK, Brazil D, Monaghan J (1999) Modelling of temperature and forces when orthogonally machining hardened steel. Int J Mach Tools Manuf 39:885–903

33. Borouchaki H, Cherouat A, Laug P, Saanouni K (2002) Adaptive re-meshing for ductile fracture prediction in metal forming. CR Mec 330:709–716

34. Vaz M Jr, Owen DRJ, Kalhori V, Lundblad M, Lindgren L-E (2007) Modelling and simulation of machining processes. Arch Comput Methods Eng 14:173–204

35. Guo YB, Dornfeld DA (2000) Finite element modeling of burr formation process in drilling 304 stainless steel. Trans ASME: J Manuf Sci Eng 122:612–619

36. Vaz M Jr, Owen DRJ (2001) Aspects of ductile fracture and adaptive mesh refinement in damaged elasto-plastic materials. Int J Numer Methods Eng 50:29–54

37. Wen Q, Guo YB, Todd BA (2006) An adaptive FEA method to predict surface quality in hard machining. J Mater Process Technol 173:21–28

38. Vaz M Jr (2000) On the numerical simulation of machining processes. J Braz Soc Mech Sci 22(2):179–188

39. Lin Z-C, Pan W-C (1993) A thermoelastic-plastic large deformation model for orthogonal cutting with tool flank wear—part I. Int J Mech Sci 35:829–840
40. Lo S-P (2000) An analysis of cutting under different rake angles using the finite element method. J Mater Process Technol 105:143–151
41. Mamalis AG, Horváth M, Branis AS, Manolakos DE (2001) Finite element simulation of chip formation in orthogonal metal cutting. J Mater Process Technol 110:19–27
42. Ng E-G, El-Wardany TI, Dumitrescu M, Elbestawi MA (2002) Physics-Based simulation of high speed machining. Mach Sci Technol 6:301–329
43. Mabrouki T, Rigal J-F (2006) A contribution to a qualitative understanding of thermo-mechanical effects during chip formation in hard turning. J Mater Process Technol 176:214–221
44. Iwata K, Osakada K, Terasaka Y (1984) Process modeling of orthogonal cutting by the rigid plastic finite element method. ASME J Eng Ind 106:132–138
45. Ceretti E, Lucchi M, Altan T (1999) FEM simulation of orthogonal cutting: serrated chip formation. J Mater Process Technol 95:17–26
46. Klocke F, Raedt H-W, Hoppe S (2001) 2D-FEM simulation of the orthogonal high speed cutting process. Mach Sci Technol 5:323–340
47. Ko D-C, Ko S-L, Kim B-M (2002) Rigid-thermoviscoplastic finite element simulation of non-steady-state orthogonal cutting. J Mater Process Technol 130–131:345–350
48. Umbrello D, Hua J, Shivpuri R (2004) Hardness-based flow stress and fracture models for numerical simulation of hard machining AISI 52100 bearing steel. Mater Sci Eng, A 374:90–100
49. Ee KC, Dillon OW Jr, Jawahir IS (2005) Finite element modeling of residual stresses in machining induced by cutting using a tool with finite edge radius. Int J Mech Sci 47:1611–1628
50. Olovsson L, Nilsson L, Simonsson K (1999) An ALE formulation for the solution of two-dimensional metal cutting problems. Comput Struct 72:497–507
51. Jaspers SPFC, Dautzenberg JH (2002) Material behaviour in metal cutting: strains, strain rates and temperatures in chip formation. J Mater Process Technol 121:123–135
52. Astakhov VP (2006) Tribology of metal cutting. Elsevier, London
53. Athavale SM, Strenkowski JS (1998) Finite element modeling of machining: from proof-of-concept to engineering applications. Mach Sci Technol 2(2):317–342
54. Usui E, Maekawa K, Shirakashi T (1981) Simulation analysis of built-up edge formation in machining low carbon steels. Bull Jpn Soc Precis Eng 15:237–242
55. Maekawa K, Shirakashi T, Usui E (1983) Flow stress of low carbon steel at high temperature and strain rate (part 2). Bull Jpn Soc Precis Eng 17(3):167–172
56. Childs THC, Otieno AMW, Maekawa K (1994) The influence of material flow properties on the machining of steels. In: Proceedings of the 3rd international conference on the behaviour of materials in machining, Warwick, pp 104–119
57. Oxley PLB (1989) The mechanics of machining: an analytical approach to assessing machinability. Ellis Horwood, Chichester
58. Johnson GR, Cook WH (1983) A constitutive model and data for metals subjected to large strains, high strain rates and high temperatures. In: Proceedings of the 7th international symposium on ballistics, The Hague, The Netherlands, pp 541–547
59. Jaspers SPFC, Dautzenberg JH (2002) Material behaviour in conditions similar to metal cutting: flow stress in the primary shear zone. J Mater Process Technol 122:322–330
60. Lee WS, Lin CF (1998) High-temperature deformation behavior of ti6al4 V alloy evaluated by high strain-rate compression tests. J Mater Process Technol 75:127–136
61. Özel T, Karpat Y (2007) Identification of constitutive material model parameters for high-strain rate metal cutting conditions using evolutionary computational algorithms. Mater Manuf Processes 22:659–667
62. Umbrello D, M'Saoubi R, Outeiro JC (2007) The influence of johnson-cook material constants on finite element simulation of machining of AISI 316L steel. Int J Mach Tools Manuf 47:462–470

63. Liang R, Khan AS (1999) A critical review of experimental results and constitutive models for BCC and FCC metals over a wide range of strain rates and temperatures. Int J Plast 15:963–980
64. Shi J, Liu CR (2004) The influence of material models on finite element simulation of machining. Trans ASME: J Manuf Sci Eng 126:849–857
65. Zerilli FJ, Armstrong RW (1987) Dislocation-mechanics-based constitutive relations for material dynamics calculations. J Appl Phys 61:1816–1825
66. Meyer HW Jr, Kleponis DS (2001) Modeling the high strain rate behavior of titanium undergoing ballistic impact and penetration. Int J Impact Eng 26:509–521
67. Madhavan V, Adibi-Sedeh AH (2005) Understanding of finite element analysis results under the framework of oxley's machining model. Mach Sci Technol 9:345–368
68. Markopoulos AP, Kantzavelos K, Galanis N, Manolakos DE (2011) 3D finite element modeling of high speed machining. Int J Manuf Mater Mech Eng 1(4):1–18
69. Zorev NN (1963) Interrelationship between shear processes occurring along tool face and on shear plane in metal cutting. In: Proceedings of the international research in production engineering conference, ASME, New York, pp 42–49
70. Childs THC, Maekawa K (1990) Computer-aided simulation and experimental studies of chip flow and tool wear in the turning of low alloy steels by cemented carbide tools. Wear 139:235–250
71. Sekhon GS, Chenot J-L (1993) Numerical simulation of continuous chip formation during non-steady orthogonal cutting. Eng Comput 10:31–48
72. Arrazola PJ, Özel T (2010) Investigations on the effects of friction modeling in finite element simulation of machining. Int J Mech Sci 52:31–42
73. Filice L, Micari F, Rizzuti S, Umbrello D (2007) A critical analysis on the friction modelling in orthogonal machining. Int J Mach Tools Manuf 47:709–714
74. Childs THC (2006) Friction modelling in metal cutting. Wear 260:310–318
75. Iqbal SA, Mativenga PT, Sheikh MA (2008) contact length prediction: mathematical models and effect of friction schemes on FEM simulation for conventional to HSM of AISI 1045 steel. Int J Mach Mach Mater 3(1/2):18–32
76. Childs THC, Maekawa K, Obikawa T, Yamane Y (2000) Metal machining: theory and applications. Elsevier, MA
77. Atkins AG (2006) Toughness and oblique cutting. Trans ASME: J Manuf Sci Eng 128(3):775–786
78. Rosa PAR, Martins PAF, Atkins AG (2007) Revisiting the fundamentals of metal cutting by means of finite elements and ductile fracture mechanics. Int J Mach Tools Manuf 47:607–617
79. Ng EG, Aspinwall DK (2002) Modeling of hard part machining. J Mater Process Technol 127:222–229
80. Usui E, Shirakashi T (1982) Mechanics of machining—from "Descriptive" to "Predictive" theory. In: Kops L, Ramalingam S (eds) on the art of cutting metals—75 Years later: a tribute to Taylor FW. In: Proceedings of the winter annual meeting of the ASME PED, vol 7, pp 13–35
81. Lin ZC, Lin SY (1992) A couple finite element model of thermo-elastic-plastic large deformation for orthogonal cutting. ASME J Eng Ind 114:218–226
82. Carrol JT III, Strenkowski JS (1988) Finite element models of orthogonal cutting with application to single point diamond turning. Int J Mech Sci 30:899–920
83. Xie JQ, Bayoumi AE, Zbib HM (1998) FEA modeling and simulation of shear localized chip formation in metal cutting. Int J Mach Tools Manuf 38:1067–1087
84. Liu CR, Guo YB (2000) Finite element analysis of the effect of sequential cuts and tool-chip friction on residual stresses in a machined layer. Int J Mech Sci 42:1069–1086
85. Shet C, Deng X (2000) Finite element analysis of the orthogonal metal cutting process. J Mater Process Technol 105:95–109

86. Obikawa T, Sasahara H, Shirakashi T, Usui E (1997) Application of computational machining method to discontinuous chip formation. Trans ASME: J Manuf Sci Eng 119:667–674
87. Komvopoulos K, Erpenbeck SA (1991) Finite element modeling of orthogonal metal cutting. ASME J Eng Ind 113:253–267
88. Lei S, Shin YC, Incropera FP (1999) Thermo-mechanical modeling of orthogonal machining process by finite element analysis. Int J Mach Tools Manuf 39:731–750
89. Zhang B, Bagchi A (1994) Finite element simulation of chip formation and comparison with machining experiment. ASME J Eng Ind 116:289–297
90. Soo SL, Aspinwall DK, Dewes RC (2004) 3D FE modelling of the cutting of inconel 718. J Mater Process Technol 150:116–123
91. Lin ZC, Lin Y-Y (1999) Fundamental modeling for oblique cutting by thermo-elastic-plastic FEM. Int J Mech Sci 41:941–965
92. Lin Z-C, Lin Y-Y (2001) Three-dimensional elastic-plastic finite element analysis for orthogonal cutting with discontinuous chip of 6–4 brass. Theoret Appl Fract Mech 35:137–153
93. Lin Z-C and Lo S-P (2001) 2-D discontinuous chip cutting model by using strain energy theory and elastic-plastic finite element method. Int J Mech Sci 43:381–398
94. Hua J, Shivpuri R (2004) Prediction of chip morphology and segmentation during the machining of titanium alloys. J Mater Process Technol 150:124–133
95. Li K, Gao X-L, Sutherland JW (2002) Finite element simulation of the orthogonal metal cutting process for qualitative understanding of the effects of crater wear on the chip formation process. J Mater Process Technol 127:309–324
96. Shet C, Deng X (2003) Residual stresses and strains in orthogonal metal cutting. Int J Mach Tools Manuf 43:573–587
97. Hashemi J, Tseng A, Chou PC (1994) Finite element modeling of segmental chip formation in high-speed machining. J Mater Eng Perform 3:712–721
98. Obikawa T, Usui E (1996) Computational machining of titanium alloy-finite element modeling and a few results. Trans ASME: J Manuf Sci Eng 118:208–215
99. Benson DJ, Okazawa S (2004) Contact in a multi-material Eulerian finite element formulation. Comput Methods Appl Mech Eng 193:4277–4298
100. Markopoulos AP (2006) Ultrprecision material removal processes. Ph.D. Thesis, National Technical University of Athens, Greece
101. Kahlori V (2001) Modelling and simulation of mechanical cutting. Ph.D. Thesis, Luleå University of Technology, Luleå, Sweden
102. Zienkiewicz OC (1971) The finite element method in engineering science. McGraw-Hill Inc, London
103. Klamecki BE (1973) Incipient chip formation in metal cutting—A 3D finite element analysis, Ph.D. Thesis, University of Illinois at Urbana Champaign
104. Shirakashi T, Usui E (1974) Simulation analysis of orthogonal metal cutting mechanism. In: Proceedings of the 1st international conference on production engineering, part I, pp 535–540
105. Dirikolu MH, Childs THC, Maekawa K (2001) Finite element simulation of chip flow in metal machining. Int J Mech Sci 43:2699–2713
106. Strenkowski JS, Moon KJ (1990) Finite element prediction of chip geometry and tool/workpiece temperature distributions in orthogonal metal cutting. ASME J Eng Ind 112:313–318
107. Strenkowski JS, Carroll JT III (1985) A finite element model of orthogonal metal cutting. ASME J Eng Ind 107:346–354
108. Özel T, Altan T (2000) Process simulation using finite element method—prediction of cutting forces, tool stresses and temperatures in high-speed flat end milling process. Int J Mach Tools Manuf 40:713–738
109. Ceretti E, Lazzaroni C, Menegardo L, Altan T (2000) Turning simulations using a three-dimentional fem code. J Mater Process Technol 98:99–103

110. Armarego EJA, Arsecularatne JA, Mathew P, Verezub S (2001) A CIRP survey on the available predictive performance models of machining operations-report on preliminary findings. In 4th CIRP international workshop on modelling of machining operations, Delft, The Netherlands, pp 002071–83

111. Mackerle J (1999) Finite-element analysis and simulation of machining: a bibliography (1976–1996). J Mater Process Technol 86:17–44

112. Mackerle J (2003) Finite element analysis and simulation of machining: an addendum a bibliography (1996–2002). Int J Mach Tools Manuf 43:103–114

113. Özel T, Zeren E (2007) Finite element modeling the influence of edge roundness on the stress and temperature fields induced by high-speed machining. Int J Adv Manuf Technol 35:255–267

114. Arrazola PJ, Ugarte D, Domínguez X (2008) A new approach for friction identification during machining through the use of finite element modelling. Int J Mach Tools Manuf 48:173–183

115. Aurich JC, Bil H (2006) 3D finite element modelling of segmented chip formation. Ann CIRP 55(1):47–50

116. Attanasio A, Ceretti E, Rizzuti S, Umbrello D, Micari F (2008) 3D finite element analysis of tool wear in machining. Ann CIRP 57(1):61–64

117. Özel T (2009) Computational modeling of 3D turning: influence of edge micro-geometry on forces, stresses, friction and tool wear in PcBN tooling. J Mater Process Technol 209:5167–5177

118. Tang DW, Wang CY, Hu YN, Song YX (2009) Finite-element simulation of conventional and high-speed peripheral milling of hardened mold steel. Metall Mater Trans A 40A:3245–3257

119. Pittalà GM, Monno M (2010) 3D finite element modeling of face milling of continuous chip material. Int J Adv Manuf Technol 47:543–555

120. Klocke F, Kratz H (2005) Advanced tool edge geometry for high precision hard turning. Ann CIRP 54(1):47–50

121. Li S, Shih AJ (2006) Finite element modeling of 3D turning of titanium. Int J Adv Manuf Technol 29:253–261

122. Mamalis AG, Kundrák J, Markopoulos A, Manolakos DE (2008) On the finite element modeling of high speed hard turning. Int J Adv Manuf Technol 38(5–6):441–446

123. Davim JP, Maranhão C, Faria P, Abrão A, Rubio JC, Silva LR (2009) Precision radial turning of AISI D2 steel. Int J Adv Manuf Technol 42:842–849

124. Markopoulos AP, Manolakos DE (2010) Finite element analysis of micromachining. J Manuf Technol Res 2(1–2):17–30

125. Bäker M (2006) Finite element simulation of high-speed cutting forces. J Mater Process Technol 176:117–126

126. Ceretti E, Fallböhmer P, Wu WT, Altan T (1996) Application of 2D FEM to chip formation in orthogonal cutting. J Mater Process Technol 59:169–180

127. Ambati R (2008) Simulation and analysis of orthogonal cutting and drilling processes using LS-DYNA. Msc. Thesis, University of Stuttgart, Germany

128. Özel T (2003) Modeling of hard part machining: effect of insert edge preparation in CBN cutting tools. J Mater Process Technol 141:284–293

129. Grzesik W (2006) Determination of temperature distribution in the cutting zone using hybrid analytical-FEM technique. Int J Mach Tools Manuf 46:651–658

130. Bil H, Kılıç SE, Tekkaya AE (2004) A comparison of orthogonal cutting data from experiments with three different finite element models. Int J Mach Tools Manuf 44:933–944

131. Childs THC, Rahmad R (2009) The effect of a yield drop on chip formation of soft carbon steels. Mach Sci Technol 13:1–17

132. Childs THC (2009) Modelling orthogonal machining of carbon steels. Part I: strain hardening yield delay effects. Int J Mech Sci 51:402–411

133. Childs THC, Rahmad R (2009) Modelling orthogonal machining of carbon steels. Part II: comparisons with experiments. Int J Mech Sci 51:465–472

Chapter 4
Application of FEM in Metal Cutting

4.1 Questions and Answers on the Performance of Machining FEM Models

In this chapter some examples of FEM models of metal cutting will be presented and discussed. The areas of application pertain to High Speed Machining (HSM), 3D modeling and micromachining. These areas are selected because either they are at the forefront of modern technology or at the forefront of advances in modeling. In either case, the topics discussed in Chap. 3 are incorporated into the models in order to obtain high quality simulations.

The last remark brings to mind a question: *are the results of the analysis accurate?* The complexity of the problem was treated in the previous Chapter. Metal cutting problems are non-linear, dynamic, require a stress and heat conduction analysis and depend on many parameters such as friction coefficient, cutting fluid action and material anisotropy that are rarely taken into account. Furthermore, FEM is a stepwise method; equations may be exact but the method introduces inaccuracies. Finally, errors are introduced by the modeler as well, in the code used or in the data provided. The accuracy of the model does not depend on how many digits the results provided have and the numerical results are never identical to the experimental ones. Of course the solution is more accurate with the increase of the equations but this must be kept within manageable limits. What is desired is that the predicted results exhibit a logical discrepancy from the antici-pated ones and that the model is able to provide equally reliable results for small alterations of the model parameters. The more general a model is, e.g. it can be used with various workpiece materials, the better the model is considered. Models are validated against experimental results or numerical results from other models.

Is the modeling method or the program wrong? FEM has been used in many areas with quite a success. Programs are created and used by humans, some not trained. Human error is a possibility. The utilization of a commercial FEM program diminishes this possibility but it does not obliterate it. In Chap. 3 it was

A. P. Markopoulos, *Finite Element Method in Machining Processes*,
SpringerBriefs in Manufacturing and Surface Engineering,
DOI: 10.1007/978-1-4471-4330-7_4, © The Author(s) 2013

discussed that the program may be correct but the results are not. More careful modeling may be required.

What can one do to get the maximum of a FEM cutting model? Incorporate into the model some, if not all, of the parameters discussed. However, the parameters still remain controversial and not a generally accepted method is proposed.

Finally, *are there any success stories in FEM simulation of metal cutting?* Yes, some of them are reviewed and discussed in the next paragraphs.

4.2 High Speed Machining Modeling

High Speed Machining is of special interest to the industry and the academia in the last few years due to the advantages it exhibits in comparison to conventional machining. The boundary between conventional machining and HSM depends on factors such as the workpiece material and the process. Most commonly, definitions of HSM make use of cutting speeds pertaining to turning, separately for ferrous and non-ferrous materials; the limit above which an operation is characterized as HSM for some non-ferrous materials can be higher at about one order of magnitude to that of alloyed steel. However, the highest cutting speeds can be achieved for non-ferrous materials that exhibit good machinability, such as aluminium, but they are limited by the attained cutting speeds of the machine tools. On the other hand, machining speeds of materials with poor machinability, such as titanium, are limited by the available cutting tools. Furthermore, operations such as turning, milling and grinding are more suitable for performing HSM than other operations, based on the achievable cutting speeds of each type of machining operation [1]. Thus, definitions that account only for cutting speed or only one cutting operation or wide material groups tend to have a lot of exceptions and to soon be outdated due to the ongoing research regarding machine tools and cutting tools for HSM.

Considering the above, a global definition of HSM operations is rather difficult to be provided since a number of factors need to be accounted for; cutting speed, spindle speed, feed, the cutting operation, workpiece material, cutting tool and cutting forces are the features included in some definitions [1–3], while a definition including the cutting tool and spindle dynamics has been proposed [4]. A definition by Tlusty [5] states that HSM refers to processes with cutting speed or spindle rotational speed substantially higher than some years before or also than the still common and general practice. The definition, even though is very general, avoids to give a spectrum of speeds or a lower value above which a machining process is characterized as HSM and can be applied to various materials and processes.

Although the rate of tool wear increases at high speeds [6], thus reducing tool life, several features of HSM can be considered as advantageous. A very important advantage of HSM is the high material removal rate achieved, which is a function of the cutting speed as well as the undeformed chip cross-section, and leads to higher productivity, especially in the case of light metal alloys. Besides the increase in the material removal rate an increase in surface quality is achieved with HSM, thus making these processes suitable for precision machining and micromachining; high speed milling is used for the fabrication of tools e.g. EDM electrodes and dies, while high speed drilling is used for micro-drills on printed circuits. The excellent surface finish reported in HSM operations, further reduces machining time and cost as it makes subsequent finishing operations, such as grinding, redundant.

It is understood that when cutting speed is increased, a subsequent increase in cutting temperatures takes place and no decrease is observed with further increase in speed, despite the predictions of the opposite phenomenon by some researchers. On the other hand with increased speed, a decrease in cutting forces is observed [7]. As experimental work has shown, cutting forces tend to reduce, in some cases by 10–15 %, as the speed is increased to high values [2, 8–10]. This force reduction may be attributed to the reduced strength of the workpiece material due to the elevated temperatures of the process [8]. High temperatures in HSM may be observed when cutting fluids are reduced or even omitted and so is their cooling effect. Cutting fluid cost, impact on the environment and inability to cool and lubricate in limited time contact needs to be taken under consideration [11, 12]. It is evident that dry machining would be preferable but tool wear does not allow this in the modern production environment. Other explanations for the force reduction are either a decrease in friction or the tendency of many materials to produce segmented chip at high cutting speeds; this saw-toothed chip also referred as "shear localized" is considered by some researchers to be energetically favorable and thus resulting in lower cutting forces [13, 14]. In any case, lower loads simplify part fixture design and allow for the machining of thin-walled sections, a common geometry of workpieces in the aerospace industry.

A special case is High Speed Hard Turning (HSHT); hard turning is a machining operation used for the processing of hardened steels which employs cubic boron nitrite (CBN) cutting tools. These are advanced cutting tools with exquisite properties, even at elevated temperatures, allowing for their application at high cutting speeds, even without the use of any cutting fluids [15]. Hard turning is providing a lot of advantages and is used in numerous applications; in today's industry it is considered as an alternative for a variety of processes such as grinding and electrical discharge machining (EDM) due to the reduced machining time required, offering accuracy equal to or better than that provided so far, with considerable cost reduction [16–19]. With HSHT great time reduction in processing can be achieved.

In FEM modeling literature regarding HSM, Marusich and Ortiz [20] were among the first to provide a model of HSM and simulate the segmented chip formation; the transition from continuous to segmented chip with increasing cutting tool speed is accomplished. The proposed model is an explicit Lagrangian model of orthogonal cutting. In the publication there is a thorough description of the features of the model in regard to mesh on mesh contact with friction and full thermo-mechanical coupling, which is realized with the so-called staggered procedure. The deformation induced element distortion is overcome with continuous re-meshing and adaptive meshing and mesh smoothing algorithms. Besides mesh distortion adaptive meshing is used to refine the contact regions along with a mesh coarsening algorithm in the inactive areas so that the problem is not too big. A fracture model is used in order to arbitrarily initiate and propagate a crack on the chip so that segmented chip is simulated. This model evolved to become AdvantEdge software, also used for simulating HSHT [21].

Bäker [13] proposed an orthogonal machining model that implements a generic flow stress law in order to simulate the cutting force reduction and the chip formation. Hortig and Svendsen [22] investigated the dependence of element size and orientation on chip formation, also using adaptive mesh refinement. Machining of aluminium alloys under high speeds is the main objective of a work presented by Davim et al. [23]. Iqbal, Mativenga and Sheikh [24] provide FEM models based on variable Coulomb and hybrid sticking-sliding friction models. Finally, some researchers utilize the Johnson–Cook material model and fracture criterion for modeling the HSM of hardened steel [14, 25].

Umbrello offers an interesting investigation on the simulation of HSM of Ti6Al4 V alloy [26]. Titanium is a difficult to machine material due to its low thermal conductivity and chemical reactivity with cutting tools, which leads to fast tool wear. However, its attractive properties, high strength for low density and corrosion resistance, make it suitable for various applications in aerospace and medical sector. Titanium alloys, under certain conditions produce segmented chips. In the investigation discussed, a plane-strain orthogonal coupled thermo-mechanical model is constructed with commercial general purpose FEM software DEFORM-2D. The thermo-viscoplastic behavior of the Titanium alloy is modeled through the Johnson–Cook model. Three different sets for the material constants of the model are obtained from the literature. Chip breakage is realized by the Cockroft-Latham criterion, which is expressed as:

$$D = \int_0^{\varepsilon_f} \sigma_1 d\varepsilon \qquad (4.1)$$

When the integral of the maximum normal stress σ_1 over the plastic strain path reaches damage value D, fracture occurs; this is the onset of breakage on the chip. The element that has reached this value is deleted and its rough boundaries are smoothed. The friction model adopted here is that of the constant shear, see Sect. 3.2.4, where the shear yield stress is [27]:

$$\tau_o = \frac{\sigma_o}{\sqrt{3}} \qquad (4.2)$$

In Fig. 4.1 the predicted segmented chip morphology for the three different sets of the Johnson–Cook material model versus experimental chip profile are depicted. It is obvious that the three chips are quite different. The author explains that when shear localization of the chip in HSM is non-linearly and dynamically modeled, an irregular shape is due to the competition between shear strain hardening and

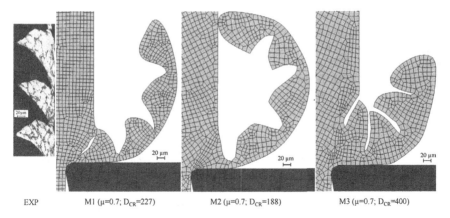

EXP   M1 (μ=0.7; D_{CR}=227)   M2 (μ=0.7; D_{CR}=188)   M3 (μ=0.7; D_{CR}=400)

Fig. 4.1 Experimental and predicted chip when machining Ti6Al4 V with cutting speed 4,800 m/min and feed 0.07 mm/rev and for three different sets of Johnson–Cook material model constants [26]

thermal softening. In the present analysis, other cutting speeds are considered, too; it is concluded that the shape of the chip is influenced by the cutting speed.

4.3 3D Machining Modeling

Models and simulations of 3D nature are not so common and the reasons behind this are that for 3D modeling and simulation the degree of complexity and the computational power required are increased. Three-dimensional models provide more detailed information of stresses, strains and temperatures within the workpiece and the tool and the chip curl of the chip. They are more realistic than orthogonal models, as will be discussed. 3D FEM modeling was used in order to simulate orthogonal and oblique cutting conditions in turning of aluminum and steel [28]. Hard tuning was modeled with the aid of 3D models [29–31]. Models that simulated segmented chip and models for tool wear are reported [32–34]. Furthermore, 3D models of milling [35, 36] and drilling [37] are presented.

In the analysis presented here, high speed turning experiments are carried out first in order to determine the cutting conditions to be simulated with FEM [38]. Four cutting parameters are taken into account. i.e. tool type, depth of cut, feed and cutting speed. For efficiency in the design of experiments, orthogonal arrays are employed. This way a full factorial analysis, which is rather laborious, is avoided; instead a fractional factorial analysis is used. The value of fractional factorial experiments in general lies in the fact that higher order interactions are usually negligible. This leads to a notable reduction in the number of parameters that need to be considered in the analysis of the data from such experiments. This, in turn, also leads to a reduction in the number of treatment combinations to be used in an experiment and hence to a reduction in the number of observations to be taken.

Orthogonal arrays are the foundation for design of experiments in Taguchi methodology and are capable of providing useful data for a small amount of experiments. In all the carried out experiments the processes took place in an OKUMA LB10ii CNC revolver turning machine with a maximum spindle speed of 10,000 rpm and a 10 HP drive motor. The workpiece, a bar of C45 cold drawn, is a common steel used in industry; during the experiments cutting forces are measured.

The obtained results are then used for the 3D simulation of HSM. The provided models are 3D turning models developed with AdvantEdge software, which integrates special features appropriate for machining simulation. It allows the simulation of various manufacturing processes such as turning, drilling, milling and micromachining among others, in either two or three dimensions. AdvantEdge is a Lagrangian, explicit, dynamic code which can perform coupled thermo-mechanical transient analysis. The program applies adaptive meshing and continuous remeshing for chip and workpiece, allowing for accurate results. The program menus are properly designed so that model preparation time is minimized. Furthermore, it possesses a wide database of workpiece and tool materials commonly used in cutting operations, offering all the required data for effective material modeling. The commercial FEM code employed makes the implementation of the latest developments very easy, reduces the model construction time and enhances the reliability of the models.

Workpiece material, cutting tools and the processes' setup are modelled from the software menus and data library, with minimum intervention from the user. This may be considered as a drawback in some cases. Nevertheless, in most cases the defaults of the program can handle the machining operation simulation adequately.

The constitutive model of the workpiece material adopted in the analysis is governed by the Power Law described by the following equation:

$$\sigma(\varepsilon, \dot{\varepsilon}, T) = g(\varepsilon) \cdot \Gamma(\dot{\varepsilon}) \cdot \Theta(T) \tag{4.3}$$

where $g(\varepsilon)$ is strain hardening, $\Gamma(\dot{\varepsilon})$ is strain rate sensitivity and $\Theta(T)$ is thermal softening. The strain hardening function $g(\varepsilon)$ is defined as:

$$g(\varepsilon) = \sigma_o \left(1 + \frac{\varepsilon}{\varepsilon_o} \right)^{1/n}, if \quad \varepsilon < \varepsilon_{cut} \tag{4.4}$$

$$g(\varepsilon) = \sigma_o \left(1 + \frac{\varepsilon_{cut}}{\varepsilon_o} \right)^{1/n}, if \quad \varepsilon \geq \varepsilon_{cut} \tag{4.5}$$

with σ_0 the initial yield stress, ε is the plastic strain, ε_o is the reference plastic strain, ε_{cut} is the cut-off strain and n is the strain hardening exponent.

The rate sensitivity function $\Gamma(\dot{\varepsilon})$ is provided as:

$$\Gamma(\dot{\varepsilon}) = \left(1 + \frac{\dot{\varepsilon}}{\dot{\varepsilon}_0}\right)^{\frac{1}{m_1}}, if \dot{\varepsilon} \leq \dot{\varepsilon}_t \tag{4.6}$$

$$\Gamma(\dot{\varepsilon}) = \left(1 + \frac{\dot{\varepsilon}}{\dot{\varepsilon}_0}\right)^{\frac{1}{m_2}} \left(1 + \frac{\dot{\varepsilon}_t}{\dot{\varepsilon}_0}\right)^{\left(\frac{1}{m_1} - \frac{1}{m_2}\right)} if \dot{\varepsilon} > \dot{\varepsilon}_t \tag{4.7}$$

where $\dot{\varepsilon}$ is strain rate, $\dot{\varepsilon}_0$ is reference plastic strain rate, $\dot{\varepsilon}_t$ is strain rate where the transition between low and high strain rate sensitivity occurs, m_1 is the low strain rate sensitivity coefficient, m_2 is the high strain rate sensitivity coefficient.

The thermal softening function $\Theta(T)$ is defined as:

$$\Theta(T) = c_0 + c_1 T + c_2 T^2 + c_3 T^3 + c_4 T^4 + c_5 T^5 \ if \ T < T_{cut} \tag{4.8}$$

$$\Theta(T) = \Theta(T_{cut}) - \frac{T - T_{cut}}{T_{melt} - T_{cut}} \ if \ T \geq T_{cut} \tag{4.9}$$

where c_0 through c_5 are coefficients for the polynomial fit, T is the temperature, T_{cut} is the linear cut-off temperature and T_{melt} is the melting temperature.

All the required data for the workpiece material used for the analysis where taken from the material database of the software. The cutting tool is modelled as rigid body. AdvantEdge allows for up to three coating layers for the cutting tools. For the analysis a tool with three layers, namely TiN, Al_2O_3 and TiC was considered. The coating layers are important for the thermal analysis that is also of interest in the present study, besides the cutting forces. The predicted cutting forces from the FEM model are compared with the measured ones, while a comparison of the chips produced in each case is also provided. Additionally, the simulations can provide more results such as the temperature fields on the cutting tool and within the workpiece. Finally, the employed software incorporates Coulomb friction across the rake face in order to model the friction at the tool-chip interface.

In Fig. 4.2 the initial set-up of the model, a snapshot of the analysis, temperatures on the tool tip and the workpiece and a real and a simulated chip can be seen. Comparison of experimental to numerical results shows good agreement.

4.4 FEM Modeling of Micromachining

Micromachining has established itself as a very important microfabrication process, over the past decade. This is attributed to the fact that, compared to other micro-fabrication processes, such as non-traditional machining and grinding, it can provide complex shapes in a wide variety of materials [1]. Furthermore, micromachining has proved to offer high quality at relatively low cost, leading to its implementation to even more industrial sectors where microfabrication is required. Micro- and recently nanomachining are in the forefront of advancements in areas such as the IT related components manufacturing, health and biomedicine, automotive industry and

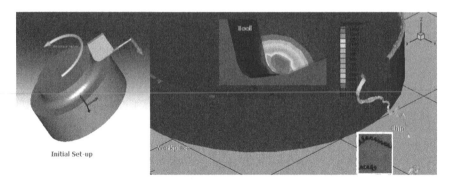

Fig. 4.2 3D model of steel machining with depth of cut 0.4 mm, feed 0.3 mm/rev and cutting speed 600 m/min

telecommunications, just to name some. Their widespread is also owed to the fact that they are incorporated into MEMS technology, where mechanical and optical microproducts are integrated with electronic parts. Mechanical and optical parts of miniature dimensions are of significant importance in MEMS while only a small number of MEMS rely solely on electronics [39].

Because of the importance of micromachining a lot of effort is dedicated to its theoretical and experimental study. Many modeling and simulation techniques have been applied so far in microtechology in general and in micromachining in particular [40], and of course FEM is one of them. However, in the micro-scale some considerations need to be considered that differentiate the "macro" from the "micro" regime. For instance, the assumption of a perfectly sharp cutting tool is non-realistic when micromachining is studied. In metal cutting, size effect, the non-linear increase in the specific energy and thus in the specific cutting force with decreasing depth of cut, influences process parameters, e.g. the minimum cutting edge radius, and therefore the analysis of the size effect is very important.

Over the years, several explanations on size effect have been reported. Defects contained in metals such as grain boundaries and impurities that form discontinuous microcracks play significant role in small dimensions. These cracks are usually formed on the primary shear plane but because of the compressive stress tend to weld and reform as strain evolves; the probability to find stress reducing defects in the shear plane is thus reduced [41]. Other explanations attribute the size effect to the relative increase of friction energy, to the heat distribution in the cutting area or to hardening effects due to strain gradients [42, 43].

Other researchers believe that the depth of cut is responsible for the size effect in micromachining [44, 45]. In micromachining the depth of cut is similar to the tool edge radius and significant sliding across the clearance face of the tool due to elastic recovery of the workpiece material is observed. Additionally, plowing due to the tool edge radius that presents the tool with a large effective negative rake angle is involved in the process. Analytical modeling indicated that the size effect

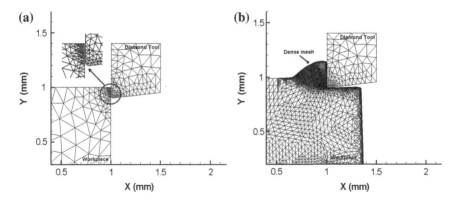

Fig. 4.3 **a** Initial mesh and **b** Mesh after a few increments

in micromachining and the cutting tool geometry plays an important role and needs to be taken into account [46].

Figure 4.3 shows a FEM model of micromachining prepared with AdvantEdge. In this figure, the continuous meshing and the adaptive remeshing procedures can be observed at the simulation progress. In Fig. 4.3a, in the related detail, the mesh of the workpiece is denser near the tool tip, where deformation is about to take place. The mesh coarsens in the areas with certain distance from workpiece surface. In the diamond tool, the mesh is denser near the tip, where more information will be acquired during the analysis. Figure 4.3b clearly indicates that new elements are created in the chip formation zone, where the strain rate is expected to be high; finer mesh can follow the curve of the curling material more closely and, furthermore, provide more accurate results.

Improvement of the material model used in micromachining is quite important; if a model that accounts for microstructure changes during micromachining is implemented in the code the obtained results are expected to be more accurate. Furthermore, all the above mentioned FEM models refer to isotropic materials; no crystallographic effects are considered in the modeling process. However, FEM simulations of the micromachining of heterogeneous materials exist [47, 48].

References

1. Byrne G, Dornfeld D, Denkena B (2003) Advancing cutting technology. Ann CIRP 52(2):483–507
2. Erdel BP (2003) High-speed machining. Society of Manufacturing Engineers, Michigan
3. Grzesik W (2008) Advanced machining processes of metallic materials: theory, modelling and applications. Elsevier, Oxford
4. Smith S, Tlusty J (1997) Current trends in high-speed machining. transactions of the ASME. J Manuf Sci Eng 119:664–666
5. Tlusty J (1993) High-speed machining. Ann CIRP 42(2):733–738

6. List G, Sutter G, Bi XF (2009) Investigation of tool wear in high speed machining by using a ballistic set-up. Wear 267:1673–1679
7. Schulz H, Moriwaki T (1992) High-speed machining. Ann CIRP 41(2):637–643
8. Trent EM, Wright PK (2000) Metal cutting. Butterworth-Heinemann, Woburn
9. Schulz H (2001) Scientific fundamentals of HSC. Carl Hanser Verlag, Munich
10. Grzesik W (2002) Developments in metal removal processes. Proceedings of the 4th International Scientific Conference "Development of Metal Cutting", Kosice, Slovakia: 103–110
11. Nouari M, Ginting A (2004) Wear characteristics and performance of multi-layer CVD-coated alloyed carbide tool in dry end milling of titanium alloy. Surf Coat Technol 200(18–19):5663–5676
12. Mamalis AG, Kundrák J, Markopoulos A, Manolakos DE (2008) On the finite element modeling of high speed hard turning. Int J Adv Manuf Technol 38(5–6):441–446
13. Bäker M (2006) Finite element simulation of high-speed cutting forces. J Mater Process Technol 176:117–126
14. Tang DW, Wang CY, Hu YN, Song YX (2009) Finite-element simulation of conventional and high-speed peripheral milling of hardened mold steel. Metall Mater Trans A 40A:3245–3257
15. Tönshoff HK, Arendt C, Ben Amor R (2000) Cutting of hardened steel. Ann CIRP, 49(2):547–566
16. Klocke F, Eisenblätter G (1997) Dry cutting. Ann CIRP 46(2):519–526
17. Kundrák J, Mamalis AG, Markopoulos A (2004) Finishing of hardened boreholes: grinding or hard cutting? Mater Manuf Proc 19(6):979–993
18. Kundrák J (2004) Applicability of hard cutting for machining of hardened bore-holes. Proceedings of the TMCE 2004, Lausanne, Switzerland: 649–660
19. Bartarya G, Choudhury SK (2012) State of the art in hard turning. Int J Mach Tools Manuf 53:1–14
20. Marusich TD, Ortiz M (1995) Modelling and simulation of high-speed machining. Int J Numer Meth Eng 38:3675–3694
21. Mamalis AG, Markopoulos AP, Kundrák J (2009) Simulation of high speed hard turning using the finite element method. J Mach Form Technol 1(1/2):1–16
22. Hortig C, Svendsen B (2007) Simulation of chip formation during high-speed cutting. J Mater Process Technol 186:66–76
23. Davim JP, Maranhão C, Jackson MJ, Cabral G, Grácio J (2008) FEM Analysis in high speed machining of aluminium alloy (Al7075-0) using polycrystalline diamond (PCD) and cemented carbide (K10) cutting tools. Int J Adv Manuf Technol 39:1093–1100
24. Iqbal SA, Mativenga PT, Sheikh MA (2008) Contact length prediction: mathematical models and effect of friction schemes on FEM simulation for conventional to HSM of AISI 1045 steel. Int J Mach Mach Mater 3(1/2):18–32
25. Duan CZ, Dou T, Cai YJ, Li YY (2009) Finite element simulation and experiment of chip formation process during high speed machining of AISI 1045 hardened steel. Int J Recent Trends Eng 1(5):46–50
26. Umbrello D (2008) Finite element simulation of conventional and high speed machining of Ti6Al4 V alloy. J Mater Process Technol 196:79–87
27. Filice L, Micari F, Rizzuti S, Umbrello D (2007) A critical analysis on the friction modelling in orthogonal machining. Int J Mach Tools Manuf 47:709–714
28. Ceretti E, Lazzaroni C, Menegardo L, Altan T (2000) Turning simulations using a three-dimensional FEM code. J Mater Process Technol 98:99–103
29. Guo YB, Liu CR (2002) 3D FEA modeling of hard turning. ASME J Manuf Sci Eng 124:189–199
30. Klocke F, Kratz H (2005) Advanced tool edge geometry for high precision hard turning. Ann CIRP 54(1):47–50
31. Arrazola PJ, Özel T (2008) Numerical modelling of 3-D hard turning using arbitrary eulerian lagrangian finite element method. Int J Mach Mach Mater 3:238–249

32. Aurich JC, Bil H (2006) 3D Finite element modelling of segmented chip formation. Ann CIRP 55(1):47–50
33. Attanasio A, Ceretti E, Rizzuti S, Umbrello D, Micari F (2008) 3D Finite element analysis of tool wear in machining. Ann CIRP 57(1):61–64
34. Özel T (2009) Computational modeling of 3D turning: influence of edge micro-geometry on forces, stresses, friction and tool wear in PcBN tooling. J Mater Process Technol 209:5167–5177
35. Bacaria J-L, Dalverny O, Caperaa S (2001) A three dimensional transient numerical model of milling. Proceedings of the Institution of Mechanical Engineers, Part B: Journal of Engineering Manufacture, 215(B8):1147–1150
36. Soo SL (2003) 3D Modelling when high speed end milling inconel 718 superalloy. Ph.D. Thesis, University of Birmingham, UK
37. Guo YB, Dornfeld DA (2000) Finite element modeling of burr formation process in drilling 304 stainless steel. Trans ASME—J Manuf Sci Eng 122:612–619
38. Markopoulos AP, Kantzavelos K, Galanis N, Manolakos DE (2011) 3D Finite element modeling of high speed machining. Int J Manuf, Mater Mech Eng 1(4):1–18
39. Jackson MJ (2007) Micro and nanomanufacturing. Springer, New York
40. Markopoulos AP, Manolakos DE (2009) Modeling and simulation techniques used in micro and nanotechnology and manufacturing. Micro Nanosystems 1:105–115
41. Shaw MC (2003) The size effect in metal cutting. Sadhana—Acad Proc Eng Sci 28(5): 875–896
42. Dinesh D, Swaminathan S, Chandrasekar S, Farris TN (2001) An intrinsic size-effect in machining due to the strain gradient. ASME/MED-IMECE 12:197–204
43. Joshi SS, Melkote SN (2004) An explanation for the size-effect in machining using strain gradient plasticity. Trans ASME—J Manuf Sci Eng 126:679–684
44. Lucca DA, Rhorer RL, Komanduri R (1991) Energy dissipation in the ultraprecision machining of Copper. Ann CIRP 40(1):69–72
45. Kim KW, Lee WY, Sin HC (1999) A finite element analysis for the characteristics of temperature and stress in micro-machining considering the size effect. Int J Mach Tools Manuf 39:1507–1524
46. Kim JD, Kim DS (1996) On the size effect of micro-cutting force in ultraprecision machining. JSME Int J, Ser C 39(1):164–169
47. Chuzhoy L, DeVor RE, Kapoor SG (2003) Machining simulation of ductile iron and its constituents, part 2: numerical simulation and experimental validation of machining. Trans ASME—J Manuf Sci Eng 125:192–201
48. Park S, Kapoor SG, DeVor RE (2004) Mechanistic cutting process calibration via microstructure-level finite element simulation model. Trans ASME—J Manuf Sci Eng 126:706–709

Chapter 5
Other Machining Processes and Modeling Techniques

5.1 Other Machining Processes

In this chapter, other machining processes, except the ones already analyzed in the first four chapters of this book, are considered. First, grinding, an abrasive process, which is the most widely used of its kind in industry, is analyzed. Modeling of grinding with FEM is quite different from modeling of turning, milling or drilling; this is why it is chosen to be analyzed individually. Furthermore, a few remarks on modeling with FEM of non-conventional machining process are made.

5.1.1 Grinding

Grinding is a precision material removal process usually used as a finishing operation. The cutting tool of the process is the grinding wheel. The grinding wheel is a bonded abrasive tool; it consists of abrasive elements hold together by a bonding material. Material removal is obtained by the interaction of the grains with the workpiece surface. Grinding exhibits similarities with orthogonal cutting when each grain of the grinding wheel is considered; the principles of a wedged tool are attributed to the grain. For more details on the processes the following books are recommended [1, 2].

As a manufacturing process, grinding is able to produce high workpiece surface quality. Improvements in its performance have allowed for the use of grinding in bulk removal of metal, maintaining at the same time its characteristic to be able to perform precision processing, thus opening new areas of application in today's industrial practice. The ability of the process to be applied on metals and other difficult-to-machine materials such as ceramics and composites is certainly an advantage of this manufacturing method. However, the energy per unit volume of material being removed from the workpiece during grinding is very large. This

A. P. Markopoulos, *Finite Element Method in Machining Processes*,
SpringerBriefs in Manufacturing and Surface Engineering,
DOI: 10.1007/978-1-4471-4330-7_5, © The Author(s) 2013

energy is almost entirely converted into heat, causing a significant rise of the workpiece temperature and, therefore, thermal damage. The areas of the workpiece that are affected are described as heat affected zones. Thermal load is connected to the maximum workpiece temperature reached during the process and therefore the maximum temperature of the ground workpiece surface is of great importance. Nevertheless, certain difficulties arise when measuring surface temperatures during grinding, mainly due to the set-up of the process; a lot of research pertaining to grinding is performed through modeling and simulation instead of experimental investigation. The importance of heat transfer phenomena over the mechanical is the reason for the existence of more thermal models in grinding than any other kind of modeling, as will be discussed in the next paragraph.

5.1.2 Modeling of Grinding

A collection of grinding models and simulation can be found in [3]; analytical, kinematic, physical/empirical, finite elements, molecular dynamics and artificial intelligence models are considered. In this survey it can be clearly seen that publications pertaining to grinding are increasing. The authors state that from the early 1970s until 2004, some 30,000 publications on grinding are cited on-line while more than 2,500 of them refer to modeling and simulation of grinding, FEM being quite an important part of them, especially in recent years. Grinding models with FEM are also cited in [4, 5], exhibiting an increase in the use of this method in the last years. In a more recent review, grinding FEM models are divided into macro- and micro-scale models to describe whether the action of the grinding wheel as a whole or the action at the level of an individual grain is considered [6]. Most of the models cited refer to two-dimensional thermal models. Grinding models are used for the prediction of surface roughness, wear characteristics, grinding forces, grinding energy and surface integrity among others. Grinding forces are essential for calculating grinding energy, which in turn determines surface integrity; grinding energy is transformed into heat dissipated into wheel, chip, workpiece and cutting fluid, if present. Excessive heat loading of the workpiece leads to the formation of heat affected zones. This heat input is responsible for a number of defects in the workpiece like metallurgical alterations, microcracks and residual stresses. High surface temperatures are connected to these phenomena and may lead to grinding burn [7, 8]. Thermal models relate all the process parameters in order to determine grinding temperatures.

Almost all thermal models of grinding are based on the moving heat sources model suggested by Jaeger [9]. In Jaeger's model the grinding wheel is represented by a heat source moving along the surface of the workpiece with a speed equal to the workspeed, see Fig. 5.1. The heat source is characterized by a physical quantity, the heat flux, q, that represents the heat entering the workpiece per unit time and area and it is considered to be of the same density along its length, taken equal to the geometrical contact length, l_c. The contact length can be geometrically

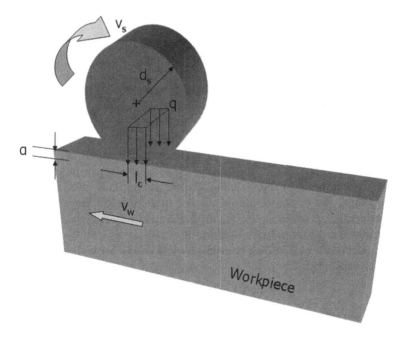

Fig. 5.1 Jaeger's model applied in grinding

calculated, assuming that deformations and motions can be neglected for depth of cut, α, a lot smaller than the grinding wheel diameter, d_s, as it is realistic in most cases in grinding, by the following equation:

$$l_c = \sqrt{a \cdot d} \tag{5.1}$$

Early papers treated moving heat source models analytically [10–12]. Other studies were performed in order to determine the energy partition between grinding wheel and workpiece [13–15] while some early FEM modeling attempts can be found in the works of Snoeys et al. [16] and Tönshoff et al. [17]. Most of the FEM thermal models that can be found in the literature pertain to 2D models with rectangular heat source profile [18, 19]. However, other source profiles such as triangular [20] can be found in the literature. All the models described refer to shallow grinding and no chip is modeled; attempts to simulate material removal are scarce [21]. Some models reported take into account the effect of material properties dependence on temperature and cooling through convection boundary conditions [22] and the use of 3D models is also present in the literature [23, 24]. Finally, coupled thermo-mechanical models are provided by researchers; these models are more complete than thermal ones in the sense that stresses from the interaction of the wheel with the workpiece are also calculated [25]. In micro-scale modeling of grinding, orthogonal cutting principles are adopted to simulate the grain as a wedge indenting the workpiece [26], or simulate 2D and 3D scratching test, with a single grain passing through the workpiece [27].

5.1.3 A FEM Grinding Model

In this paragraph a thermal grinding model is presented. It is realised by the commercial finite element software MSC.Marc Mentat. Jaeger's model is incorporated to the program. Furthermore, some special features that improve the model are included, providing a novel, more efficient and reliable simulation of precision grinding, e.g. the two coefficients of the workpiece material that are related to temperature, i.e. the thermal conductivity and the specific heat capacity, are considered to be temperature depended. Transient conditions and temperature depended material properties produce non-linear finite element problems, which are more difficult to be solved. Additionally, the cutting fluid effect is simulated; this is a feature that is not taken into account in the original Jaeger's model but adds considerably to the accuracy and the building of a sound model of grinding.

The mathematical formulation used for heat transfer analysis by MSC.Marc is concisely given below. The heat transfer problem can be written, as known, as a differential equation:

$$[C]\{\dot{T}\} + [K]\{T\} = \{Q\} \tag{5.2}$$

$[C]$ is the heat capacity matrix, $[K]$ the conductivity and convection matrix, $\{T\}$ the vector of the nodal temperatures and $\{Q\}$ the vector of nodal fluxes. In the case of a steady state problem, where $\dot{T} = \frac{\partial T}{\partial t} = 0$, the solution can be easily obtained by a matrix inversion:

$$\{T\} = [K]^{-1}\{Q\} \tag{5.3}$$

In the case of transient analysis, where $\dot{T} \neq 0$, which is the case described here, the nodal temperature is approximated at discrete points in time as:

$$\{T\}^n = \{T\}(t_0 + n\Delta t) \tag{5.4}$$

MSC.Marc is using a backward difference scheme to approximate the time derivative of the temperature:

$$\{\dot{T}\}^n \cong \frac{\{T\}^n - \{T\}^{n-1}}{\Delta t} \tag{5.5}$$

which results in the finite difference scheme:

$$\left(\frac{[C]}{\Delta t} + [K]\right)\{T\}^n - \frac{[C]}{\Delta t}\{T\}^{n-1} = \{Q\} \tag{5.6}$$

that gives the solution of the differential Eq. 5.2.

For the models, based on Jaeger's moving source theory, the heat flux q needs to be determined. The heat flux can be calculated from the following equation:

$$q = \varepsilon\frac{F'_t \cdot v_s}{l_c} \tag{5.7}$$

where ε is the percentage of heat flux entering the workpiece, F'_t the tangential force per unit width of the workpiece, v_s the peripheral wheel speed and l_c the contact length. The proportion of the heat flux entering the workpiece can be calculated by a formula suggested by Malkin [7] for grinding with aluminum oxide wheels, by making assumptions on the partitioning of total specific grinding energy, u, required for grinding. The total specific grinding energy consists of three different components: the specific energy required for the formation and the removal of the chip, u_{ch}, the specific energy required for plowing, i.e. the plastic deformation in the regions where the grains penetrate the workpiece surface but no material is removed, u_{pl} and the specific energy required for making the flat wear grains slide on the workpiece surface, u_{sl}, thus:

$$u = u_{ch} + u_{pl} + u_{sl} \tag{5.8}$$

It has been analytically and experimentally shown that approximately 55 % of the chip formation energy and all the plowing and sliding energy are conducted as heat into the workpiece, i.e.

$$\varepsilon = \frac{0.55 \cdot u_{ch} + u_{pl} + u_{sl}}{u} = \frac{u - 0.45 \cdot u_{ch}}{u} \Rightarrow \varepsilon = 1 - 0.45 \frac{u_{ch}}{u} \tag{5.9}$$

The component u_{ch} has a constant value of about 13.8 J/mm^3 for grinding for all ferrous materials. The total specific grinding energy is calculated from the following equation:

$$u = \frac{F'_t \cdot v_s}{a \cdot v_w} \tag{5.10}$$

where v_w is the workspeed and, consequently, as in Jaeger's model, the speed of the moving heat source. Note that, in both Eqs. 5.7 and 5.10 the value of F'_t is needed in order to calculate the heat flux and the total specific grinding energy, respectively; it can be calculated from the power per unit width of the workpiece, P'_t, as follows:

$$F'_t = \frac{P'_t}{v_s} \tag{5.11}$$

The last equation suggests that if the power per unit width of the workpiece is known, then the heat flux can be calculated. In order to provide the appropriate data for FEM models, i.e. the heat flux, the tangential force per unit width of the workpiece and surface temperatures, the power per unit width of the workpiece needs to be measured. This can be realized by experimental work [28]. Six aluminum oxide grinding wheels of the same diameter $d_s = 250$ mm and width $b_s = 20$ mm with different bonding were used on a BRH 20 surface grinder. Four depths of cut were used, namely 10, 20, 30 and 50 μm while the workpiece speed was $v_w = 8$ m/min and the wheel speed $v_s = 28$ m/s kept constant for all sets of experiments, for all wheels. The workpiece materials were the 100Cr6, C45 and X210Cr12 steels. Throughout the process the synthetic coolant Syntilo-4 was applied at 15 l/min. For each grinding wheel, 10 passes of the same depth of cut were performed over the workpiece. The power per unit width of the workpiece was measured for each pass and its average value was calculated. For measuring the power, a precision three-phase wattmeter was used. First, the power of the idle grinding machine was measured and set as the zero point of the instrument. Then, the workpieces were properly ground and the power was registered on the measuring device. After 10 passes were performed the grinding wheel was dressed with a single point diamond dressing tool, with depth $a_d = 0.02$ mm and feed of $f_d = 0.1$–0.2 mm/wheel rev. In total, 72 measurements took place.

The boundary conditions of the finite element model are applied; on the top surface heat is entering the workpiece in the form of heat flux that moves along the surface. Cooling from the applied cutting fluid is simulated by means of convective boundary conditions. All the other sides of the workpiece are considered to be adiabatic, and so no heat exchange takes place in these sides. The cooling effect simulated refers to the flood method, where coolant at low pressure and room temperature fills the upper part of the workpiece, applying a uniform cooling in all the surface area.

The model needs to have a sufficient enough length in order for the temperature fields to be deployed and observed in full length. A mesh, consisting of four-noded rectangular full integration elements with one degree of freedom, namely the temperature, for the thermal models, is applied on the workpiece geometry for plane stress analysis. The mesh is denser towards the grinding surface, which is the thermally loaded surface, and, thus the most affected zone of the workpiece, allowing for greater accuracy to be obtained; this is realized in the same basis as the discretization of the primary and secondary deformation zone in orthogonal cutting, i.e. the mesh is denser where more results are needed. The mesh is refined only in the vertical direction for two reasons. First of all, the elements of the top row, representing the ground surface, have the same dimensions so that the boundary conditions representing a uniform heat flux sliding across the workpiece surface can be modeled. Secondly, the presented model is a thermal model, with no mechanical interactions and thus no deformation that would require local refinement. Vertical refinement allows the user to observe the phenomena on the workpiece surface in detail without requiring extra computer time. Thermal modeling presented here has been validated and used before [29]; it has proved to work well, predicting grinding temperatures with accuracy.

Figure 5.2 presents the temperature contours within a workpiece of material 100Cr6. The heat input causes the rapid increase of temperature. The maximum temperature varies between different cutting conditions, material and cutting wheel.

In Fig. 5.3 the temperature variation on the workpiece surface for workpiece material 100Cr6 and depth of cut 50 μm, for all six wheels is presented. From the figure it can be concluded that the temperature fields appear to be the same for the same depth of cut and the only difference is the maximum temperature reached for each grinding wheel. Furthermore, it is revealed that the temperatures are higher in

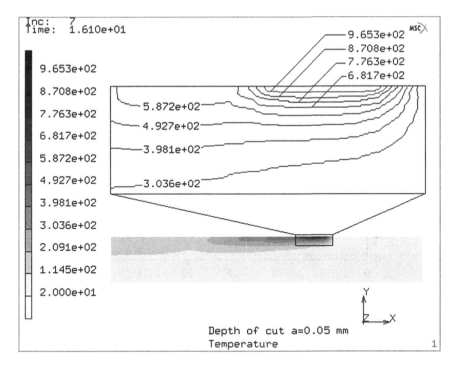

Fig. 5.2 Temperature contours within the workpiece

the regions on the back of the wheel; therefore, it seems that, it is more critical to direct the coolant to this side, in order to prevent the damage of the surface integrity due to the temperature rise. Maximum temperatures vary from 600° to more than 900 °C, depending on the grinding wheel used. Such values for the maximum temperature, when grinding steels, are reported by other investigators, too [15, 16, 30, 31].

The high temperatures that appear in grinding have a negative effect on the workpiece. The surface of the workpiece and also the layers that are near the surface and have been affected by the heat loading during the grinding process consist the heat affected zones of the workpiece. The excessive temperature in these zones contributes to residual stresses, microcracking and tempering and may cause microstructure changes, which result to hardness variations of the workpiece surface. Steels that cool down quickly from temperatures above the austenitic transformation temperature undergo metallurgical transformations; as a result, untempered martensite is produced in the workpiece. Excessive heat may also lead to metallurgical burn of the workpiece, which produces a bluish color on the surface of the processed material due to oxidation. If the critical temperatures at which these transformations take place are known, the size of the heat affected zones can be also predicted from the FEM model. The actual size of these zones and their composition depends on the duration of thermal loading, except the maximum

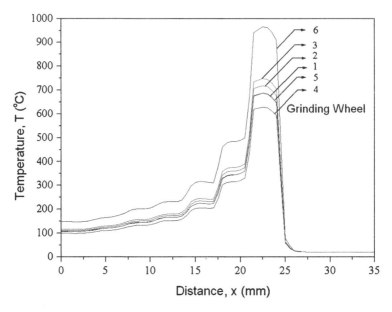

Fig. 5.3 Temperature variation on the surface of the workpiece for depth of cut 50 μm and workpiece material 100Cr6

temperature reached. The three critical temperatures for the 100Cr6 steel are $T_t = 150$ °C for tempering, $T_m = 250$ °C for martensitic and, $T_a = 800$ °C for austenitic transformation and are related to hardness variation, residual stresses and the formation of untempered martensite layers within the workpiece [14, 32–34].

In Fig. 5.4 the variation of the temperatures within the workpiece with depth below the surface is shown, as calculated for all grinding wheels, for 100Cr6 for depth of cut 50 μm. These temperatures are taken underneath the grinding wheel where the maximum temperatures are reached. In the same diagram the three critical temperatures for the 100Cr6 steel are also indicated. From these diagrams the theoretical depth of the heat affected zones, for each wheel used and depth of cut can be determined. When grinding with grinding wheel 6, austenitic transformation temperature is exceeded in the layers with depth up to 0.1 mm below the surface. There is the possibility, when machining hardened steels, to exceed critical temperatures that may damage the workpiece and create heat affected zones through the metallurgical transformations that take place [32]. This may be attributed to the use of unsuitable grinding wheel, non-effective cooling or inappropriate grinding conditions. On the other hand, it should be taken into account that the top surface of the workpiece, that is the part of the workpiece mostly affected by the thermal damage, is to some extent, depending on the depth of cut, carried away as a chip.

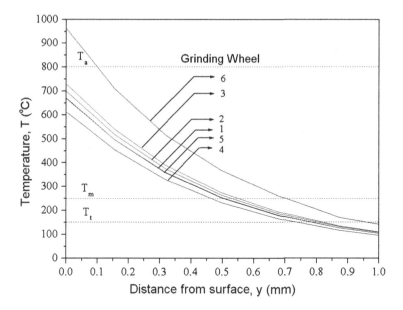

Fig. 5.4 Variation of temperature versus distance from surface when grinding 100Cr6 steel with all grinding wheels for depth of cut 50 μm

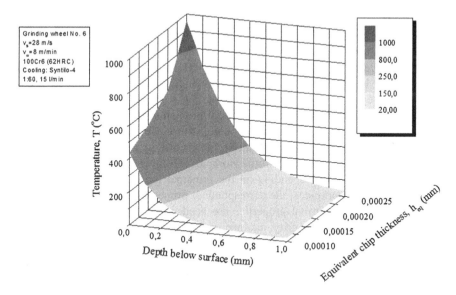

Fig. 5.5 Variation of temperature with depth below surface and equivalent chip thickness, h_{eq}, when grinding 100Cr6 steel with grinding wheel 6

Figure 5.5 shows the temperatures on the surface and within the workpiece when using grinding wheel 6 on 100Cr6, for different values of the equivalent chip thickness, h_{eq} (mm), which is calculated as:

$$h_{eq} = \frac{v_w}{v_s} \alpha \qquad (5.12)$$

The equivalent chip thickness includes the effect of three grinding parameters and is more suitable than the depth of cut to be used for the optimization of the grinding conditions. In order to observe, or to limit, its critical value it is not necessarily needed to decrease the depth of cut; it is also possible to alter suitably the grinding conditions. In the same diagram the regions within the critical temperatures are also indicated by contour bands, so that the heat affected zones can be also predicted. Such diagrams can be constructed for other materials as well and used as a guide for choosing the optimal grinding conditions. Note, however, that the critical temperatures are not the same for all steels.

5.1.4 Non-Conventional Machining

Non-conventional machining processes include Electro-Discharge Machining (EDM), Water Jet Machining (WJM), Laser Machining (LM), Electro-Chemical Machining (ECM) among others. The difference of these processes with conventional ones is that there is no mechanical interaction between a cutting tool and the workpiece. As in the case of cutting, these processes need to be investigated for fundamental understanding and for producing predictive models for their performance. Usually, the material removal principle is quite different from one another. For every process, a different modeling strategy needs to be followed. Finite element method can treat all physical problems and with the right formulation, models for these processes can be constructed. In [4, 5, 27, 35, 36] FEM models for several such processes can be found. The inputs of these models are mainly material properties, both of the workpiece from which material is removed but also from the tool. Depending on the principal used by the tool to remove material from the workpiece, the corresponding properties are of interest, e.g. for EDM electrical and thermal properties are needed. The predictions are usually connected with surface integrity, attained under various machining conditions. In [35] material models for non-conventional machining processes are proposed.

A topic of special interest is machining of composite materials [37]. Although, usually these materials are machined with conventional processes, the models used to simulate the awkward nature of the material are quite interesting, e.g. the laser assisted machining of an alumina fiber reinforced aluminum metal matrix composite (MMC) presented in [38]. There are two methods to model these materials, namely with the use of an equivalent homogenous material or a micromechanics approach [39]. In the first technique the composite material is modeled as a homogenous material with the properties of the matrix and the fiber combined, while in the second technique the matrix and the fibers are treated separately. This area is of interest in many sectors of contemporary industry, e.g. automotive, aerospace and medical sectors and able FEM models are anticipated with great interest.

5.2 Other Modeling Methods

Although this book is dedicated to finite elements, it would not be complete without reference to some other modeling techniques that are also used in metal cutting. In the next few paragraphs some techniques, other than FEM will be briefly discussed. References will be provided for those interested to gain more details on these modeling methods.

5.2.1 Soft Computing in Machining

Soft computing techniques such as Artificial Neural Networks (ANNs), Genetic Algorithms (GA) and Fuzzy Logic (FL) are gaining more attention from researchers dealing with machining processes. This is mainly attributed to their ability to handle complex problems with a relatively easy "computational" way. As an example ANNs can simulate machining and provide predictions without taking into account any underlying physical phenomena. The model is a "black box" that is trained to provide accurate results within the limits of input that it was trained with, no matter how non-linear or multi-dimensional a problem is. Furthermore, soft computing techniques are characterized by the fact that can provide results very quick, making them suitable for on-line optimization of machining processes. The use of soft computing techniques on metal cutting and grinding is reported in [3, 40]. However, this technique is also applied to non-conventional machining, e.g. EDM and WJM [41, 42]. In a literature review by Chandrasekaran et al. [43] several soft computing techniques, namely neural networks, fuzzy sets, genetic algorithms, simulated annealing, ant colony optimization and particle swarm optimization are discussed, and their application to turning, milling, drilling and grinding is documented. The assessment of their predictive performance on various parameters such as cutting forces, tool wear and surface finish as well as their optimization performance is presented. In the next lines an ANNs model of EDM will be presented, to exhibit the potential benefits from employing this method.

An artificial neural network is defined as "a data processing system consisting a large number of simple, highly interconnected processing elements (artificial neurons) in an architecture inspired by the structure of the cerebral cortex of the brain" [44]. Actually, ANNs are models intended to imitate some functions of the human brain using its certain basic structures. ANNs have been shown to be effective as computational processors for various associative recall, classification, data compression, combinational problem solving, adaptive control, modeling and forecasting, multisensor data fusion and noise filtering [45]. Two main and important features of neural networks are their architecture, i.e., the way that the network is structured, and the algorithm used for its training. After the appropriate training, the selected network has the ability to interconnect one value of output to one particular value of input which is given.

The "core" element of a neural network is the neuron. Neurons are connected to each other with a set of links, called synapses and each synapse is described by a synaptic weight. Neurons are placed in layers and each layer's neurons operate in parallel. The first layer is the input layer. The activity of input units represents the non-processed information that entered the network and at that layer neurons do not perform any computations. The hidden layers follow the input layer. The activity of each hidden unit is determined from the activity of the input units and the weights at the connections of input and hidden units. A network can have many or none hidden layers and their role is to improve the network's performance. The existence of these layers at the network becomes more necessary as the number of input neurons grows. The last layer is the output layer. The behavior of output units depends upon the activity of the hidden units and the weights between hidden units and output units. The output of the layer is the output of the whole network; output layer neurons in contrast to input layer ones perform calculations.

There are two types of neural networks: the feed-forward and the recurrent ones. Feed-forward neural networks allow the signals to travel in only one direction: from input to output, i.e. the output signal of a neuron is the input to the neurons of the following layer and never the opposite. The inputs of the first layer are considered the input signals of the whole network and the output of the network is the output signals of last layer's neurons. On the contrary, recurrent networks include feedback loops allowing signals to travel forward and/or backward [46]. Feed-forward neural networks are characterized by simple structure and easy mathematical description [47].

In general, there is not a standard algorithm for calculating the proper number of hidden layers and neurons. For relatively simple systems, as the present case, a trial-and-error approach is usually applied in order to determine which architecture is optimal for a problem. Networks that have more than one hidden layers have the ability to perform more complicated calculations. However, for most applications, a hidden layer is enough, while for more complicated applications the simulation usually takes place using two hidden layers. The existence of more than necessary hidden layers complicates the network, resulting in a low speed of convergence during training and large error during operation. Therefore, the architecture of a neural network always depends upon the specific situation examined and must not be more complex than needed [45].

Once the number of layers and the number of units in each layer are selected the network's weights must be set in order to minimize the prediction error made by the network; this is the role of the training algorithms. The historical cases that were gathered are used to automatically adjust the weights in order to minimize this error. The error of a particular configuration of the network can be determined by running all the training cases through the network and comparing the actual output generated with the desired or target outputs. The differences are combined together by an error function resulting the network's error. Usually the mean square error (MSE) of the network's response to a vector p, is calculated, according to the equation:

$$E_p = \frac{1}{2} \sum_{i=1}^{l} \left(d_{p,i} - o_{p,i} \right)^2 \tag{5.13}$$

where $o_{p,i}$ are the values of the output vector which occur for the input vector p and $d_{p,i}$ the values of the desirable response corresponding to p. The procedure is repeated until MSE becomes zero. Each time that the program passes through all pairs of training vectors an epoch is completed; training usually ends after reaching a great number of epochs.

One of the frequently used training algorithms is the back-propagation (BP) algorithm. It is usually applied in feed-forward networks with one or more hidden layers [48]. The input values vectors and the corresponding desirable output values vectors, are used for the training of the network until a function is approached which relates the input vectors with the particular output vectors. When the value of the mean square error is calculated, it is propagated to the back in order to minimize the error with the appropriate modification of the weights.

Another important parameter of the neural network models is their ability to generalize. Generalization is the ability of neural networks to provide logic responses for input values that were not included in the training. Correctly trained back-propagation networks are able to perform generalization; this ability provides the opportunity of training the network using a representative set of input—desirable output values pairs.

When an algorithm is applied to the network random values are given to the weight factors. The convergence speed and the reliability of the network depend upon the initial values of weights; thus different results may be observed during the application of the same algorithm to the network. There are only a few elements that can guide the user for the selection of the proper values. A wrong choice may result to small convergence speed or even to network's paralysis, where the training stops. Furthermore, due to the nature of the algorithm that searches for the minimum error, the network may be stabilized at a local minimum instead of the total minimum. That results most of the times in wrong response values of the network. To overcome these problems variations of the algorithm have been created; for further information on this topic Refs. [46, 47] may be consulted. Worth mentioning, also, that a very common and simple technique used for overcoming problems of this type is the repetition of the algorithm many times and the use of different initial values of the weight factors.

One of the problems that occur during the training of neural networks is the over-fitting which undermines their generalization ability. The error appears very small at the set of the training vectors, however, when new data are imported to the network the error is becoming extremely large. This phenomenon is attributed to the fact that the network memorized the training examples; on the other hand did not learn to generalize under the new situations. The generalization ability of a network is assured when the number of training data is quite greater than the number of network's parameters. However, when the network is large the relations between input and output become rather complicated. Hence, a network should not be larger than needed to solve the given problem. Note, also, that two improvement techniques may be applied during modeling; namely, normalization of the used data and the early stopping technique.

Normalization is a method used in neural networks so that all the data present a logical correlation; all input and output data are suitably transformed so that their mean value becomes equal to zero and the standard deviation equal to one. Otherwise, the neural network could suppose that a value is more significant than the others because its arithmetic value is greater. This could damage the generalization ability of the network and lead to overfitting. After normalization all inputs are equally significant for the training of the network. For the improvement of generalization of a neural network the early stopping technique is usually employed. By this methodology the existing data are separated in three subsets. The first subset consists of the training vectors, which are used to calculate the gradient and to form the weight factors and the bias. The second subset is the validation group. The error in that group is observed during training and like training group normally decreases during the initial phase of training. However, when the network begins to adjust the data more than needed, the error in that group raises and when that increase is continued for a certain number of repetitions, training stops. Finally, the third subset is the test group and its error is not used during training. It is used to compare the different models and algorithms.

The analysis presented here pertains to the prediction of surface roughness of several electro-discharge machined steels under various conditions. Electrical discharge machining (EDM) is a thermal process with a complex metal removal mechanism, involving the formation of a plasma channel between the tool and workpiece electrodes, melting and evaporation action and shock waves, resulting in phase changes, tensile residual stresses, cracking and metallurgical transformation. These properties determine the operational behavior of machined parts. As far as EDM is concerned, the relative literature includes publications where ANNs are applied, mainly, for the estimation/prediction of the material removal rate, the optimization and the on-line monitoring of the process [49–52] whilst prediction of surface finish is presented only in [53]. The ANNs models developed take into consideration the workpiece material, the pulse current and the pulse duration as input parameters in order to predict the center-line average (R_a) surface roughness. The suggested neural networks are trained with experimental data [54]. For the formulation of the ANNs and the simulation of EDM Matlab was employed. Matlab is a well known program used for simulation purposes. Its toolbox which is exclusively used for neural networks is user-friendly and the creation of neural networks is easy using a small amount of

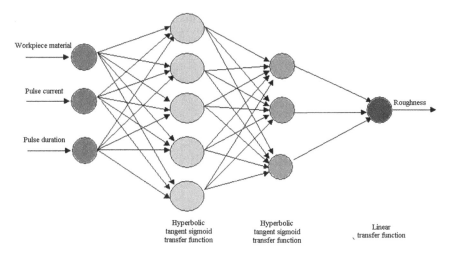

Fig. 5.6 Neural network architecture

commands; the program has a data base with functions, algorithms and commands for that purpose.

As a first step, several models were designed and tested in order to determine the optimal architecture, the most suitable activation functions and the best training algorithm suitable for the prediction of R_a. Each model was tested more than once in order to evaluate whether it truly converges to a low value or not. After this trial-and-error procedure the model selected is a feed-forward neural network with two hidden layers consisting of five and three neurons respectively. The activation function in both the hidden layers is the hyperbolic tangent sigmoid transfer function and in the output layer is the linear transfer function. The training algorithm used is the back propagation (BP) algorithm. The architecture of the selected (optimized) network is presented in Fig. 5.6. In order to use early stopping technique, ½ of the available experimental data are used for training, ¼ are used for validation and ¼ are used for testing. The selection of the data constituting the three groups is performed in stochastic way so that training is not performed partially, for example for only one workpiece material, a fact that could had lead to an erroneous generalization; moreover all data are equally represented.

For the selected ANNs model, the MSE of training is about 0.088 and its training took almost 700 epochs to complete. The MSE of all the three groups of the early stopping technique is presented in Fig. 5.7. From this figure it is evident that validation and testing group MSEs are higher than that of the training group, as expected. Moreover, they have similar values which indicate that the proposed neural network possesses good generalization ability, thus being able to model EDM process. For the evaluation of the generalization ability of the trained neural network a linear fit between the output of the model and the experimental data, for all the measured values, without discrimination to which group they belong, is performed. The linear fit is presented in Fig. 5.8; note that T and A represent the

Fig. 5.7 Results on neural
networks training

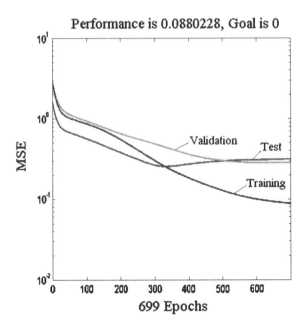

experimental results and the outputs of the model, respectively. The best linear fit function is calculated as being: $A = 0.922T + 0.934$, while the correlation coefficient is $R = 0.904$. The model can be saved and used for the prediction of surface roughness, given that the pulse current and duration are within the limits of the model and the workpiece material is one of the five steel grades tested. ANNs produce reliable results and in a timely manner. However, FEM results are richer. In order to select one method or the other, one must be aware of this fact. Hybrid models combining FEM and ANN are reported [28].

5.2.2 Molecular Dynamics

Unlike FEM and soft computing techniques that can be employed for a wide range of processes modeling, Molecular Dynamics (MD) is used for simulating nanometric cutting. It is true that FEM is a popular simulation technique for micromachining as well. However, FEM is based on principles of continuum mechanics and at nanometric level this is considered a drawback. On the other hand, Molecular Dynamics can simulate the behavior of materials in atomic scale. A review on MD simulation of machining at the atomic level can be found in [55]. Applications of MD simulation for grinding are also popular in order to study the interaction of a single grain with the workpiece [3].

MD is a modeling method in which atoms and molecules are interacting for a period of time, by means of a computer simulation. In order to simulate molecular

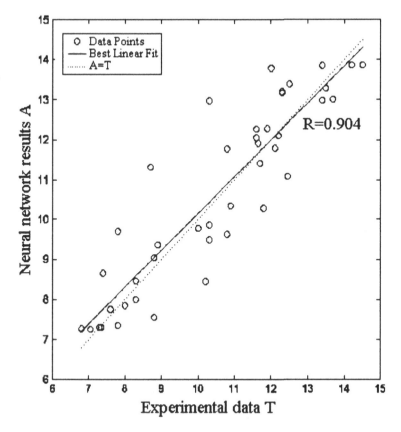

Fig. 5.8 Correlation between experimental data and neural network output

systems, a very big number of particles is involved and a vast number of equations is produced to describe the properties of these systems; as a multidisciplinary method, laws and theories from mathematics, physics and chemistry consist the backbone of the method. In order to deal with these problems, numerical methods, rather than analytical ones, are used and algorithms from computer science and information theory are employed. Although the method was originally intended to be exploited in theoretical physics, nowadays it is mostly applied in biomolecules, materials science and nanomanufacturing.

MD method was introduced in the simulation of micro and nanomanufacturing in the early 1990s [56, 57]. The results indicated that MD is a possible modeling tool for the microcutting process; atomistic modeling can provide better representation of micro and nanolevel characteristics than other modeling techniques. MD models developed were used for the investigation of the chip removal mechanisms, tool geometry optimization, cutting force estimations, subsurface damage identification, burr formation, surface roughness and surface integrity

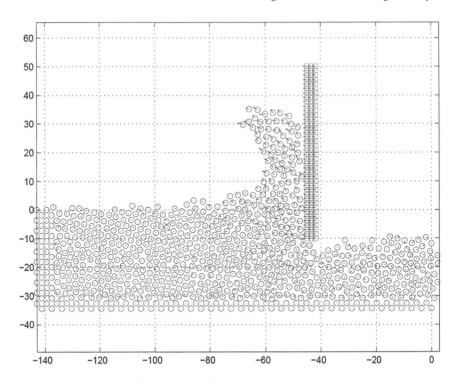

Fig. 5.9 MD simulation of nanometric cutting

prediction; some relevant works can be found in Refs. [58–69]. Figure 5.9 shows a typical MD simulation where the chip formation can be observed.

Some disadvantages can be identified in the MD technique as well. MD simulation is based on calculations of interatomic forces among a vast number of atoms that constitute the modeled system. This task requires significant computational power and in order to overcome this problem very small model sizes are simulated; some models are limited to nanometer or Angstrom level. Another feature is that cutting speed is considered to be unrealistically high, of the order of 200–500 m/s when typical speeds range between 2 and 10 m/s in microcutting, in order to bring cutting speed closer to the atomic movement speed and thus save computational time.

However, the detailed insight in material behavior in microcutting provided by MD simulation has supported process development and optimization and at the same time has provided information on its theoretical study. It is generally accepted that chip formation in cutting is owed to the shearing effect in the cutting zone of the workpiece. When machining with a depth of cut that is of the same order to the cutting edge radius, another phenomenon needs to be considered. Regardless of the nominal rake angle of the cutting tool, the effective rake angle in nanometric cutting is always negative. Thus, a compressive stress imposes deformation in front of the cutting edge. It can be concluded that under certain circumstances there is no chip formation but elastic and plastic deformation.

The subsurface deformation is also described by a MD model in [70] where nanometric cutting is performed by an AFM pin tool, a technique proposed for the fabrication of MEMS and NEMS. Additionally, models have been proposed that combine MD and FEM modeling techniques in order to exploit the capabilities of each method and cover a more wide range of material behavior at nanoscale cutting [71, 72].

5.2.3   Other Modeling Methods

It is true that the dominating numerical method used in modeling machining is the finite element method. However, other numerical methods used in modeling machining are found in the literature. The finite difference method (FDM) has been used to solve the heat transfer problem in cutting [73]. However, this numerical method is not suitable for problems where the mesh is severely distorted, as in machining, due to the reliance of the method on its mesh. Mesh update can be performed but it reduces accuracy and needs complicated programming.

A new approach would be the use of meshless methods such as the element-free Galerkin (EFG) method and the smoothed particle hydrodynamics (SPH) method [74–76]. In fact both these methods have been employed for the modeling of machining in order to study tool wear when cutting titanium [77] and to estimate optimal machining conditions with respect to surface quality [78]. Especially SPH method in [79] is carried out using LS-DYNA software.

References

1. Malkin S (1989) Grinding technology: theory and applications of machining with abrasives. Society of Manufacturing Engineers, Dearborn
2. Klocke F, König W (2005) Fertigungsverfahren. Schleifen, Honen, Läppen. Springer, Berlin
3. Brinksmeier E, Aurich JC, Govekar E, Heinzel C, Hoffmeister H-W, Klocke F, Peters J, Rentsch R, Stephenson DJ, Uhlmann E, Weinert K, Wittmann M (2006) Advances in modeling and simulation of grinding processes. Ann CIRP 55(2):667–696
4. Mackerle J (1999) Finite-element analysis and simulation of machining: a bibliography (1976–1996). J Mater Process Technol 86:17–44
5. Mackerle J (2003) Finite element analysis and simulation of machining: an addendum a bibliography (1996–2002). Int J Mach Tools Manuf 43:103–114
6. Doman DA, Warkentin A, Bauer R (2009) Finite element modeling approaches in grinding. Int J Mach Tools Manuf 49:109–116
7. Malkin S (1978) Burning limit for surface and cylindrical grinding of steels. Ann CIRP 27(1):233–236
8. Malkin S, Guo C (2007) Thermal analysis of grinding. Ann CIRP 56(2):760–782
9. Jaeger JC (1942) Moving sources of heat and the temperature at sliding contacts. J Proc R Soc N S W 76(3):203–224
10. Outwater JO, Shaw MC (1952) Surface temperature in grinding. Trans ASME 74:73–86

11. Des Ruisseaux NR, Zerkle RD (1970) Thermal analysis of the grinding process. Trans ASME J Eng Ind 92:428–434
12. Malkin S (1974) Thermal aspects of grinding: part 2 surface temperatures and workpiece burn. Trans ASME J Eng Ind 96:1184–1191
13. Lavine AS (1988) A simple model for convective cooling during the grinding process. Trans ASME J Eng Ind 110:1–6
14. Rowe WB, Petit JA, Boyle A, Moruzzi JL (1988) Avoidance of thermal damage in grinding and prediction of the damage threshold. Ann CIRP 37(1):327–330
15. Kato T, Fujii H (2000) Temperature measurement of workpieces in conventional surface grinding. Trans ASME J Manuf Sci Eng 122:297–303
16. Snoeys R, Maris M, Peters J (1978) Thermally induced damage in grinding. Ann CIRP 27(2):571–581
17. Tönshoff HK, Peters J, Inasaki I, Paul T (1992) Modelling and simulation of grinding processes. Ann CIRP 41(2):677–688
18. Mamalis AG, Kundrák J, Manolakos DE, Gyáni K, Markopoulos A, Horváth M (2003) Effect of the workpiece material on the heat affected zones during grinding: a numerical simulation. Int J Adv Manuf Technol 22:761–767
19. Biermann D, Schneider M (1997) Modeling and simulation of workpiece temperature in grinding by finite element analysis. Mach Sci Technol 1:173–183
20. Mahdi M, Zhang L (1995) The finite element thermal analysis of grinding processes by ADINA. Comput Struct 56:313–320
21. Weber T (1999) Simulation of grinding by means of the finite element analysis. Proceedings of the 3rd international machining & grinding SME conference, Ohio, USA
22. Mamalis AG, Kundrak J, Manolakos DE, Gyani K, Markopoulos A (2003) Thermal modelling of surface grinding using implicit finite element techniques. Int J Adv Manuf Technol 21:929–934
23. Jin T, Stephenson DJ (1999) Three dimensional finite element simulation of transient heat transfer in high efficiency deep grinding. Ann CIRP 53(1):259–262
24. Wang L, Qin Y, Liu ZC, Ge PQ, Gao W (2003) Computer simulation of a workpiece temperature field during the grinding process. Proc Inst Mech Eng Part B J Eng Manuf 217(7):953–959
25. Mahdi M, Zhang L (2000) A numerical algorithm for the full coupling of mechanical deformation, thermal deformation, and phase transformation in surface grinding. Comput Mech 26:148–156
26. Ohbuchi Y, Obikawa T (2003) Finite element modeling of chip formation in the domain of negative rake angle cutting. J Eng Mater Technol 125:324–332
27. Klocke F, Beck T, Hoppe S, Krieg T, Müller N, Nöthe T, Raedt HW, Sweeney K (2002) Examples of FEM application in manufacturing technology. J Mater Process Technol 120:450–457
28. Markopoulos AP (2011) Simulation of grinding by means of the finite element method and artificial neural networks. In: Davim JP (ed) Computational methods for optimizing manufacturing technology. IGI Global, Hershey, pp 193–218
29. Markopoulos AP (2011) Finite elements modelling and simulation of precision grinding. J Mach Form Technol 3(3/4):163–184
30. Hoffmeister H-W, Weber T (1999) Simulation of grinding by means of the finite element analysis. Third international machining & grinding SME conference, Ohio, MR99-234, USA
31. Moulik PN, Yang HTY, Chandrasekar S (2001) Simulation of stresses due to grinding. Int J Mech Sci 43:831–851
32. Shaw MC, Vyas A (1994) Heat affected zones in grinding steel. Ann CIRP 43(1):279–282
33. Zhang L, Mahdi M (1995) Applied mechanics in grinding—IV. The mechanism of grinding induced phase transformation. Int J Mach Tools Manuf 35:1397–1409
34. Chang CC, Szeri AZ (1998) A thermal analysis of grinding. Wear 216:77–86
35. Dixit US, Joshi SN, Davim JP (2011) Incorporation of material behavior in modeling of metal forming and machining processes: a review. Mater Des 32:3655–3670

36. Bhondwe KL, Yadava V, Kathiresan G (2006) Finite element prediction of material removal rate due to electro-chemical spark machining. Int J Mach Tools Manuf 46:1699–1706
37. Davim JP (ed) (2012) Machining of metal matrix composites. Springer, London
38. Dandekar C, Shin YC (2010) Laser-assisted machining of a fiber reinforced Al-2 %Cu metal matrix composite. Trans ASME J Manuf Sci Eng 132(6):061004
39. Soo SL, Aspinwall DK (2007) Developments in modeling of metal cutting processes. Proc Inst Mech Eng Part L J Mater Des Appl 221:197–211
40. van Luttervelt CA, Childs THC, Jawahir IS, Klocke F, Venuvinod PK (1998) Present situation and future trends in modelling of machining operations. Ann CIRP 47(2):587–626
41. Markopoulos A, Vaxevanidis NM, Petropoulos G, Manolakos DE (2006) Artificial neural networks modeling of surface finish in electro-discharge machining of tool steels (ESDA 2006-95609). Proceedings of ESDA 2006, 8th biennial ASME conference on engineering systems design and analysis, Torino, Italy
42. Vaxevanidis NM, Markopoulos A, Petropoulos G (2010) Artificial neural network modelling of surface quality characteristics in abrasive water jet machining of trip steel sheet. In: Davim JP (ed) Artificial intelligence in manufacturing research. Nova Science Publishers, Inc, New York
43. Chandrasekaran M, Muralidhar M, Murali Krishna C, Dixit US (2010) Application of soft computing techniques in machining performance prediction and optimization: a literature review. Int J Adv Manuf Technol 46:445–464
44. Tsoukalas LH, Uhrig RE (1997) Fuzzy and neural approaches in engineering. Wiley Interscience, New York
45. Davalo E, Naim P, Rawsthorne A (1991) Neural networks. Macmillan Education Limited, London
46. Fausset LV (1994) Fundamentals of neural networks: architectures, algorithms and applications. Prentice Hall, Upper Saddle River
47. Haykin S (1999) Neural networks: a comprehensive foundation. Prentice Hall, Upper Saddle River
48. Dini G (1997) Literature database on applications of artificial intelligence methods in manufacturing engineering. Ann CIRP 46(2):681–690
49. Kao JY, Tarng YS (1997) A neural network approach for the on-line monitoring of the electrical discharge machining process. J Mater Process Technol 69:112–119
50. Tsai K-M, Wang PJ (2001) Comparisons of neural network models on material removal rate in electrical discharge machining. J Mater Process Technol 117:111–124
51. Wang K, Gelgele HL, Wang Y, Yuan Q, Fang M (2003) A hybrid intelligent method for modelling the EDM process. Int J Mach Tools Manuf 43:995–999
52. Panda DK, Bhoi RK (2005) Artificial neural network prediction of material removal rate in electro discharge machining. Mater Manuf Process 20(4):645–672
53. Tsai K-M, Wang PJ (2001) Predictions on surface finish in electrical discharge machining based upon neural network models. Int J Mach Tools Manuf 41:1385–1403
54. Markopoulos AP, Manolakos DE, Vaxevanidis NM (2008) Artificial neural network models for the prediction of surface roughness in electrical discharge machining. J Intell Manuf 19(3):283–292
55. Komanduri R, Raff LM (2001) A review on the molecular dynamics simulation of machining at the atomic scale. Proc Inst Mech Eng Part B J Eng Manuf 215:1639–1672
56. Stowers IF, Belak JF, Lucca DA, Komanduri R, Moriwaki T, Okuda K, Ikawa N, Shimada S, Tanaka H, Dow TA, Drescher JD (1991) Molecular-dynamics simulation of the chip forming process in single crystal copper and comparison with experimental data. Proc ASPE Ann Meet 1991:100–104
57. Ikawa N, Shimada S, Tanaka H, Ohmori G (1991) Atomistic analysis of nanometric chip removal as affected by tool-work interaction in diamond turning. Ann CIRP 40(1):551–554
58. McGeough J (ed) (2002) Micromachining of engineering materials. Marcel Dekker, Inc., New York

59. Inamura T, Takezawa N, Kumaki Y (1993) Mechanics and energy dissipation in nanoscale cutting. Ann CIRP 42(1):79–82
60. Shimada S, Ikawa N, Tanaka H, Uchikoshi J (1994) Structure of micromachined surface simulated by molecular dynamics analysis. Ann CIRP 43(1):51–54
61. Rentsch R, Inasaki I (1995) Investigation of surface integrity by molecular dynamics simulation. Ann CIRP 42(1):295–298
62. Komanduri R, Chandrasekaran N, Raff LM (1998) Effect of tool geometry in nanometric cutting: a molecular dynamics simulation approach. Wear 219(1):84–97
63. Komanduri R, Chandrasekaran N, Raff LM (2001) MD simulation of exit failure in nanometric cutting. Mater Sci Eng A 311:1–12
64. Cheng K, Luo X, Ward R, Holt R (1993) Modeling and simulation of the tool wear in nanometric cutting. Wear 255:1427–1432
65. Luo X, Cheng K, Guo X, Holt R (2003) An investigation on the mechanics of nanometric cutting and the development of its test-bed. Int J Prod Res 41(7):1449–1465
66. Rentsch R (2004) Molecular dynamics simulation of micromachining of pre-machined surfaces. Proceedings of 4th Euspen international conference, Glascow, Scotland, pp 139–140
67. Fang FZ, Wu H, Liu YC (2005) Modelling and experimental investigation on nanometric cutting of monocrystalline silicon. Int J Mach Tools Manuf 45:1681–1686
68. Pei QX, Lu C, Fang FZ, Wu H (2006) Nanometric cutting of copper: a molecular dynamics study. Comput Mater Sci 37:434–441
69. Cai MB, Li XP, Rahman M (2007) Study of the temperature and stress in nanoscale ductile mode cutting of silicon using molecular dynamics simulation. J Mater Process Technol 192–193:607–612
70. Zhang JJ, Sun T, Yan YD, Liang YC, Dong S (2008) Molecular dynamics simulation of subsurface deformed layers in AFM-based nanometric cutting process. Appl Surf Sci 254:4774–4779
71. Aly MF, Ng E-G, Veldhuis SC, Elbestawi MA (2006) Prediction of cutting forces in the micro-machining of silicon using a "hybrid molecular dynamic-finite element analysis" force model. Int J Mach Tools Manuf 46:1727–1739
72. Lin Z-C, Huang J-C, Jeng Y-R (2007) 3D nano-scale cutting model for nickel material. J Mater Process Technol 192–193:27–36
73. Grzesik W, Bartoszuk M, Nieslony P (2004) Finite difference analysis of the thermal behaviour of coated tools in orthogonal cutting of steels. Int J Mach Tools Manuf 44: 1451–1462
74. Belytshko T, Krongauz Y, Organ D, Fleming M, Krysl P (1996) Meshless methods: an overview and recent developments. Comput Methods Appl Mech Eng 139:3–47
75. Liu GR (2002) Mesh free methods moving beyond finite element method. CRC, Boca Raton
76. Chen Y, James Lee J, Eskandarian A (2006) Meshless methods in solid mechanics. Springer, New York
77. Calamaz M, Limido J, Nouari M, Espinosa C, Coupard D, Salaun M, Girot F, Chieragatti R (2009) Toward a better understanding of tool wear effect through a comparison between experiments and SPH numerical modelling of machining hard materials. Int J Refract Metal Hard Mater 27(3):595–604
78. Gurgel AG, Sales WF, de Barcellos CS, Bonney J, Ezugwu EO (2006) An element-free Galerkin method approach for estimating sensitivity of machined surface parameters. Int J Mach Tools Manuf 46(12–13):1637–1642
79. Limido J, Espinosa C, Salaun M, Lacome JL (2007) SPH method applied to high speed cutting modelling. Int J Mech Sci 49(7):898–908